Freeman Laboratory Separates in Biology

Freeman Laboratory Separates are self-bound, self-contained exercises. They are 8½ inches by 11 inches in format and are punched for a three-ring notebook. Separates may be ordered in any assortment or quantity at 50¢ each. Thus, an instructor can construct an entire manual tailored to the design of his course or he can select Separates to broaden the coverage of the laboratory manual currently in use.

Order Separates by number. Students' sets are collated and packaged by the publisher before shipment to the bookstore.

Each exercise in this manual is available as a Separate (see below). On the facing page are listed Freeman Laboratory Separates in Biology.

DISSECTION OF THE FETAL PIG
by Warren F. Walker, Jr.

W. H. FREEMAN AND COMPANY
660 Market Street, San Francisco, California 94104
20 Beaumont Street, Oxford, England OX1 2NQ

THIRD EDITION

Dissection of the Fetal Pig

Warren F. Walker, Jr.

Oberlin College

ILLUSTRATED BY
Judith L. Dohm
Edna Indritz Steadman
and Pat Densmore

W. H. FREEMAN AND COMPANY
San Francisco

Printed in the United States of America

International Standard Book Number: 0–7167–1214–8

9 8 7 6 5 4 3 2

Contents

A Guide for the Teacher

Only a few remarks are needed for a set of anatomical exercises of the type herein. They relate to the specimens used and their care, to useful supplementary material, and to parts of the exercises for which the student might need special help.

Exercise 1

A. Fetal pigs can be obtained from any of the biological supply companies. Large specimens, ranging in length from 10 to 14 inches, should be ordered. Suppliers usually ship specimens in small plastic bags, to which they can be returned after an exercise. At Oberlin, we also give each student a plastic box in which to keep the specimen. If the specimens are to be used for more than a month or two, a small amount of an 8% formalin solution should be placed in the box and the specimens should be dipped periodically in a formalin or embalming solution. Embalming solution, which inhibits mold growth better than formalin, can be made according to the following formula:

Carbolic acid (melted crystals)	5 parts
Formalin (40%)	5 parts
Glycerin	5 parts
Water	85 parts

B. A slide of cat skin is described because it is more readily available than one of pig skin. The directions apply to the skin of most mammals.

C-2. In addition to mounted mammalian skeletons, a separate skull and isolated thoracic or lumbar vertebrae should be available for the study of vertebral structure.

C-3. A mounted skeleton or a limb bone of a young mammal should be available to show epiphyseal plates.

C-4. Demonstrate a bone cut longitudinally so that students can see compact and spongy bone. Slides of dried bone, thinly ground, show bone structure better than do microscopic sections of fixed bone.

Exercise 2

C to J. Muscles are described by regions of the body so that the instructor can select certain groups if time does not permit a complete dissection. It is necessary to dissect the shoulder muscle before those on the anterior part of the trunk can be studied, but other groups can be studied independently.

Fetal muscles are delicate and tear easily, and so

students should be cautioned to separate them carefully and not to pull on them.

K. Individual or demonstration slides of striated muscle should be provided. If slides of muscle as such are not available, striated muscle can be seen in a slide of the tongue or certain other organs.

Exercise 3

A-1. The salivary glands are very difficult to dissect; so the instructor may wish to have students study them using demonstration dissections. In any case, such a dissection should be available to help the student find the glands.

A-2. Much laboratory time can be saved if a demonstration of the procedure for opening the mouth and pharynx is given to groups of students.

Certain features of the head and neck can be seen particularly well in sagittal sections, and several should be on demonstration.

Vocal cords are poorly developed in a fetal pig but can be seen very clearly in a cow's larynx, which can usually be obtained from a local slaughter house. A "moo" can be produced by blowing through the trachea of a fresh preparation and appropriately manipulating the larynx.

B-1. Again, much time can be saved by giving a group demonstration of the procedure for opening the body cavity.

D-2. A demonstration should be available to help the students find the pancreatic duct.

D-4. The description of the small intestine is based on a slide of cat intestine, which is the most readily available type; it applies to the intestine of most mammals.

Exercise 4

If both arteries and veins are to be dissected by the student, doubly injected specimens should be used. Time and money can be saved, however, by giving the students singly injected specimens and providing demonstration dissections of the veins.

E. The heart of the fetal pig is large enough to show most parts of the mammalian heart clearly. If a fresh beef heart can be obtained from a local slaughterhouse, it will be possible to demonstrate the action of certain valves. Cut open the right atrium and nick one of the pulmonary arteries. Closure of the right atrioventricular valve can be shown by running water through a hose that passes through the opened atrium into the right ventricle. It is necessary to clamp off the pulmonary trunk so that the ventricle will fill. The return of arterial blood to the heart can be simulated by cutting open the left ventricle so that the caudal surface of the left atrioventricular valve can be seen and running water through a hose inserted into a nicked pulmonary vein. Closure of the aortic valve can be shown by tying a hose into the aorta and clamping the coronary arteries, some of which will have been cut. The aortic valve can be seen through the opened left ventricle.

G. If slides of blood are not available, the students can make slides of their own. Procedures are described in detail in books on microscopic technique, such as *Basic Microscopic Technique* by R. McC. Jones (Chicago, University of Chicago Press, 1966). Briefly, the students should use the following procedure. Wipe a finger with alcohol and stab it with a disposable sterile lance. Squeeze one drop of blood onto one end of a clean microscope slide that is placed on the table. Hold a second slide at about a 45° angle to the first, and back it up to the drop until the drop spreads out in the acute angle between the two slides. Then push the second slide across the first slide, spreading the blood into a thin film behind it. Dry the smear rapidly by waving it in the air.

Wright's stain is a good one to try. Cover the smear with a dozen drops or so of stain and leave it on for 1 or 2 minutes. Then add an equal amount of distilled water that has been buffered so that it is close to neutrality and leave it on for an additional 2 to 4 minutes. Holding the slide vertically, rinse it with distilled water. Blot the slide gently with a paper towel or filter paper, but do not wipe it   Let it dry thoroughly and study it. No cover slip is necessary.

Exercise 5

A. Triply injected sheep kidneys (arteries, veins, and ureter injected) can be obtained from supply houses, and they make excellent demonstrations.

B-3. A demonstration of the uterus of a pregnant sow will be needed if this section is to be done. Many parts of the female genital tract will also be very clear in such a preparation, including ovarian follicles and corpora lutea, ostium, uterine tube, and the parts of the uterus.

C. Individual or demonstration slides of a mammalian testis and ovary are needed.

Exercise 6

A. Most of the major features of the mammalian eye can be seen in the eye of the fetal pig if dissecting microscopes are used. Instructors may prefer to use larger sheep or cow eyes. In any case, dissections of some larger eyes should be demonstrated.

B. Because it is encased by cartilage and soft bone, the middle ear can be exposed exceptionally well in a fetal pig, but demonstrations of the inner ear will be needed. Certain supply houses market a dissection of a human temporal bone in which the chambers that contain the various parts of the inner ear have been carefully exposed.

C. Because much of the skull is cartilaginous, the nose is easy to dissect in a fetus. The vomeronasal organ must be looked for before the nasal septum is destroyed.

Exercise 7

A-2. Individual or demonstration slides of a cross section of a frog or mammalian spinal cord and nerve are needed.

B. The brains have not been hardened in a fetal pig, and so specially preserved sagittal sections of sheep brains should be used if at all possible. Specimens should be checked to be sure that they are cut in the sagittal plane. A larger half can easily be dissected down to the sagittal plane. If handled carefully, the specimens can be used by many laboratory sections, and for several years. One demonstration of a brain should have the dura mater intact, and another should have the pia mater injected.

If a lateral dissection is made of the pig brain (Fig. 7-3), the trigeminal, facial, vagus, and hypoglossal nerves in particular will be seen. A demonstration dissection would help the students identify them.

Preface

The third edition of the *Dissection of the Fetal Pig,* like its predecessors, has been prepared primarily to meet the need for a short set of exercises appropriate for college students studying the pig in introductory biology or zoology courses. Descriptions of a few sheep organs, such as the eye and brain, are also included because many find them more conveneint for study than comparable organs of the fetal pig. The emphasis is on gross anatomy, but directions are given for the microscopic study of the anatomy of selected organs. Some knowledge of the microscopic structure of organs adds greatly to an understanding of their gross structure and function, and such material is frequently presented in introductory courses. Enough material is included so that these exercises also meet the needs of students in certain comparative and mammalian anatomy courses. However, flexibility has been an aim, and an instructor can meet the requirements of courses at different levels by a selection of exercises or parts of exercises.

Sections in which students were having difficulty finding structures or understanding material have been rewritten in this edition. The mammalian skull is described more completely, and directions for dissecting the eye modified to apply to the adult sheep or cow eye as well as to the fetal eye. A brief discussion of the overall pattern of adult circulation has been placed at the beginning of the exercise on the circulatory system so that students will have a context in which to place their dissection of the arteries and veins. Discussions of the functions of many organs have been expanded somewhat to reinforce, at the time that a student sees a structure, the material presented in the textbook and in class. The structure of organs cannot be separated from their activities and functional relationships to other parts of the body. Finally, this edition includes brief sections on the microscopic anatomy of additional tissues and organs: skeletal muscle, blood cells, testis, ovary, and neurons as seen in a spinal cord and nerve. These sections should increase the students' understanding of the gross organs being dissected.

A Glossary of Vertebrate Anatomical Terms has been appended to this edition. Terms are defined and their classical origins given. The study of this glossary should help students become familiar with the basic word roots used in anatomical terminology and enable them to easily grasp the meanings of unfamiliar terms.

Favored technical terms are printed in **bold face** type in the exercises themselves. For the most part, they are anglicized versions of those recommended by the *Nomina Anatomica Veterinaria* (Vienna, 1972). This code, which applies insofar as possible the human terms of the *Nomina Anatomica Parisiensia* (1955) to quadrupeds, is becoming the standard for mammals other than human beings. The two codes differ primarily with respect to modifying adjectives for direction: thus the human "superior vena cava," for example, is referred to as the "cranial vena cava." In a few cases I use a familiar English word instead of the less familiar *Nomina* term (e.g., "liver" rather than "hepar"); however, in such cases, the *Nomina* term is given in *italics* because it often forms the basis for constructing familiar adjectives (e.g., "hepatic"). A few other common synonyms are given in roman type.

Because careful observations are essential for the acquisition of scientific knowledge, and for an understanding of the generalizations that derive from an anatomical study, directions for dissection are given in a way that will encourage students to take more than a superficial look at the organs, and yet the directions are specific enough to enable them to find structures with a minimum of assistance from the instructor. The labeled drawings are designed to help in this, as well as to give the student a record of his or her more important observations. Because much less laboratory time is now spent on straight anatomical work than formerly, students frequently do not have time to make their own drawings, even though the pencil is the best of aids to the eye.

Each exercise is based on a natural unit of material and can be fitted into the total laboratory program where it is most appropriate. Most exercises can be completed in a standard three-hour laboratory period, although some may take a bit more time and some less. Certain sections can easily be omitted or replaced by demonstrations, according to the objectives of the instructor.

I continue to be much indebted to Judith L. Dohm, Edna Indritz Steadman, and Pat Dinsmore for the great care and artistry with which they prepared final drawings from my sketches for the first two editions. I have been helped with the new illustrations for this edition by P. Anne Smith, an Oberlin student who worked closely with me, and again by Edna Indritz Steadman, who rendered them into their final form. Both have added greatly to the quality of this set of directions, and I am grateful to them. I am also much indebted to Patricia Mittelstadt, whose skillful editing of these exercises has eliminated many ambiguities and errors. I wish to thank those users of earlier editions who have taken the time and trouble to send comments to the publishers. Their encouragement has been most gratifying, and their criticisms have formed the basis for many of the changes in this edition. It is my hope that users of these exercises will continue to call to my attention any errors or parts in need of revision.

Warren F. Walker, Jr.
November 1979

A Note to the Student

If this is the first animal that you have dissected, you should read these remarks concerning procedures and terms of direction carefully before you begin. At the outset, a few incisions will have to be made with scalpel or scissors to open the specimen, but thereafter most of the dissecting should be done with a pair of fine forceps. Dissecting consists of carefully separating organs and picking away surrounding connective tissue to expose them clearly. Do not cut out any organs unless specifically directed to do so, but always pick away enough of the surrounding tissue to see all of an organ clearly. Do not be content with just a glimpse of it. A good dissection should reveal all of the organs clearly enough so that another person could examine the specimen and see the essential relationships of the organs and their connections with other organs without any difficulty.

During the dissections, you may want to spread the legs of the specimen and tie them to the dissecting pan, but it is not necessary to do this.

The terms *right* and *left* always refer to the speci-men's right and left. Depending on how the specimen is oriented, this may or may not correspond with your own right and left. *Lateral* refers to the side of the body or organ in question; *median,* to the center. When a structure is described as being *superficial to* some part, it means that it lies over the part referred to and nearer to the body surface. Conversely, a structure that is *deep to* some other structure lies beneath it and farther away from the body surface. Other terms for direction differ somewhat for a quadruped and for human beings, who stand erect (see the accompanying figure). For human beings, *superior* refers to the upper, or head, end of the body; *inferior,* to the lower parts of the body. The belly surface is *anterior* and the back is *posterior.* For quadrupeds, the underside of the body is *ventral* and the back is *dorsal.* Anterior and posterior are acceptable terms of direction for certain structures in a quadruped head because this part of the body corresponds in its orientation to the human head. However, because structures in other parts of the quadruped body have a different orientation, there is a tendency

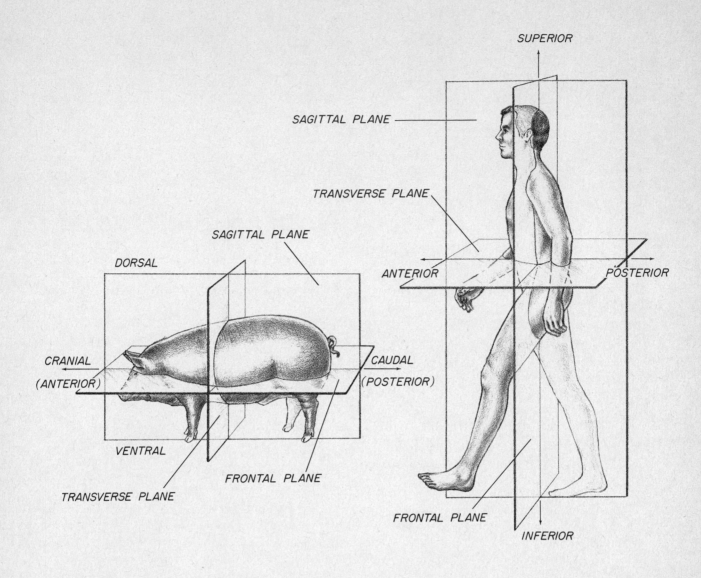

SUPERIOR

SAGITTAL PLANE

TRANSVERSE PLANE

ANTERIOR

POSTERIOR

FRONTAL PLANE

INFERIOR

SAGITTAL PLANE

DORSAL

CRANIAL
(ANTERIOR)

CAUDAL
(POSTERIOR)

VENTRAL

FRONTAL PLANE

TRANSVERSE PLANE

Planes of the body and differences in terms for direction between a quadruped and a human being.

to replace the terms anterior and posterior for most parts of the quadruped with *cranial*, for a direction toward the head, and *caudal*, for a direction toward the tail.

The *distal* end of a structure is the end farthest away from some point of reference, usually the origin of the structure or the midventral line of the body; the *proximal* end is the end nearest the point of reference.

A *sagittal section* is a section in the longitudinal plane of the body passing from the middle of the back to the middle of the belly. A *frontal section* is also a longitudinal section, but it passes from the middle of the right side to the middle of the left side. A *transverse section* crosses the longitudinal axis of the body at right angles.

At the end of a laboratory period, always clean up your work area and put your specimen away in the container provided.

REFERENCES

The references listed below will be useful to those who wish additional information on the anatomy of the pig. The list is not exhaustive—it is simply an introduction to the literature that is available.

Biological Abstracts. Philadelphia, 1926 to date. An essential bibliographic tool for those who wish to check the primary research literature. Two issues summarizing the biological research in the world are published monthly.

Evans, W. L., Lee, A. E., and Tatum, G. *The Fetal Pig: A Photographic Study*, revised edition. New York: Rinehart, 1958. A collection of labeled photographs of pig dissections.

Gilbert, S. A. *Pictorial Anatomy of the Fetal Pig*, 2nd edition. Seattle: University of Washington Press, 1966. A very well illustrated guide for the study of the pig. Includes many sketches of isolated organs and views of dissections not usually seen but helpful in understanding the relationships of organs.

International Committee on Veterinary Anatomical Nomenclature. *Nomina Anatomica Veterinaria*, 2nd edition. Vienna: Adolf Holzhausen, 1972. Gives the official terminology for veterinary anatomy, which is becoming the standard for quadruped mammals. It is distributed in the United States by the Department of Anatomy, New York State Veterinary College, Ithaca, New York 14850.

Northcutt, R. G., Williams, K. L., Barber, R. P. *Atlas of the Sheep Brain*, 2nd edition. Champaign: Stipes, 1966.

Odlaug, T. O. *Laboratory Anatomy of the Fetal Pig*, 5th edition. Dubuque: Wm. C. Brown, 1975. A guide for the study of the pig.

Sisson, S., and Grossman, J. D. In *The Anatomy of Domestic Animals*, 5th edition, edited by R. Getty. Philadelphia: Saunders, 1975. Pig structure is considered in detail in the second volume of this standard veterinary textbook.

External Anatomy, Skin, and Skeleton

A study of the anatomy of the pig (*Sus scrofa*) is particularly valuable, for the pig is more closely related to human beings than other laboratory animals commonly studied. Like human beings, fishes, frogs, turtles, birds, and pigs are backboned animals, which places them in the subphylum Vertebrata of a larger group—the phylum Chordata. The other subphyla of the chordates include such peculiar marine forms as sea squirts and lancelets. Within the vertebrates, pigs and human beings belong to the same class, the Mammalia, for they have a high level of metabolic activity (warm-bloodedness), hair, and other features to control body temperature, and the females nurse their young with milk secreted by mammary glands. Pigs are closely related to deer, sheep, and cows; all are hoofed animals having an even number of toes—four or two. Such animals constitute the order Artiodactyla. Human beings belong to the order Primates, along with lemurs, monkeys, and the great apes.

The gestation period for pigs is from 112 to 115 days; at birth they are from about 30 to 35 centimeters long. Fetal pigs used for study usually range from 25 to 35 centimeters in length.

A. EXTERNAL FEATURES

Notice that the body of the pig, like that of all higher vertebrates, consists of a **head, neck, trunk** (from which 2 pairs of **appendages** arise), and **tail.** The head (Fig. 1-1) bears the **mouth,** bounded by fleshy **lips** and **cheeks** that enable the young to suckle; **external nostrils** (*nares*) on the end of the snout; **eyes,** bounded by upper and lower **eyelids;** and the external part of the ear, which consists of an external flap, the **auricle** or pinna, and a short passage, the **external acoustic meatus,** leading to the eardrum. In most mammals these parts of the ear help gather and concentrate sound waves in the manner of an old-fashioned ear trumpet. Make an incision extending forward from the anterior corner of the eye, and pull the upper and lower eyelids apart. You will notice the **nictitating membrane** (*third eyelid*) covering the anterior part of the eyeball. This membrane can move across the eyeball, which helps to keep it clean. We have a vestige of such a membrane, the semilunar fold, in the median corner of the eye.

A mammal's trunk consists of a cranial part, the **thorax,** encased by the ribs, and a caudal part, the **abdomen.** In a fetal pig, the **umbilical cord** is attached to the ventral surface of the abdomen and connects the fetus with the placenta. Make a fresh cut across the cord about 1 centimeter from its attachment to the abdomen; notice that it contains two thick-walled **umbilical arteries,** which carry blood low in oxygen and high in waste products from the fetus to the placenta, and a larger, thin-walled **umbilical vein,** which returns blood rich in oxygen and food and low in waste products to the fetus. If the arteries of your specimen

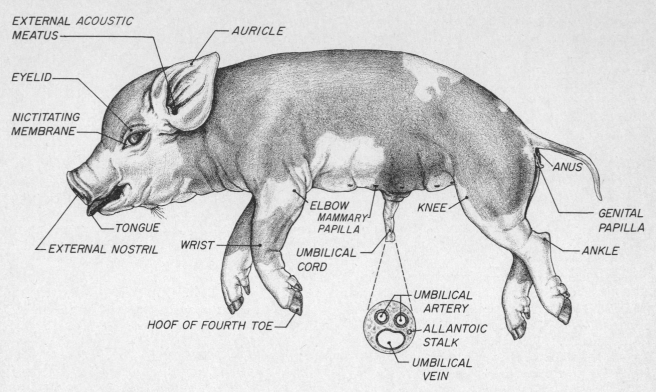

FIGURE 1-1
Lateral view of external features of a fetal pig. *Inset:* Cross section of umbilical cord, enlarged.

have been injected, there will be colored latex in them. Between, or near, the arteries you should find a small, hard cord of tissue, which is a remnant of the **allantoic stalk.** The allantois is a fetal membrane that helps to form the placenta and will be considered later (Exercise 5). All the structures within the umbilical cord are embedded in a gelatinous connective tissue.

If your specimen is a male, the **preputial orifice** of the penis will be found directly caudal to the attach-

ment of the umbilical cord (Fig. 1-2), and a sac of skin, the **scrotum,** which will contain the testes in a mature male, will be developing caudal to the hind legs. In both sexes, the caudal orifice of the digestive tract, the **anus,** is situated directly ventral to the base of the tail. In a female specimen (Figs. 1-1 and 1-2), the common orifice of the urinary tract and vagina lies directly ventral to the anus. It is bounded laterally by low folds, the **labia,** which come together ventrally to form a

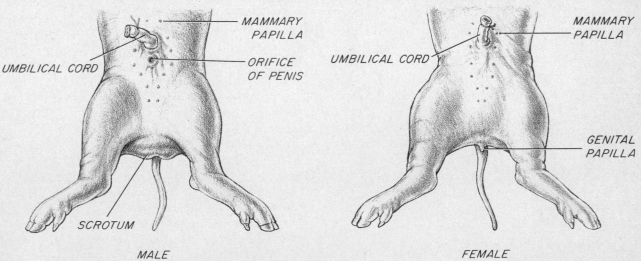

FIGURE 1-2
Ventral views of the posterior half of male and female fetal pigs, illustrating sexual dimorphism.

spikelike **genital papilla.** This region, including the orifice and the labia, is part of the vulva (Exercise 5). In both sexes, teats, or **mammary papillae,** which become part of the mammary glands in mature females, extend along the ventral surface of the trunk. The number of pairs is usually 5 or 6, ample for nursing from 6 to 12 young.

Pigs, like other ungulates, walk on the tips of their toes. The toes end in **hooves,** and there is always some reduction in the number of toes compared with the five found in most terrestrial vertebrates. In pigs, the first toe, corresponding to our thumb and great toe, has been lost. Those corresponding to our third and fourth are larger than those corresponding to our second and fifth and carry most of the body weight. Because the animal walks on its toe tips, the foot is elongated, and the **wrist** and **ankle** are carried well off the ground. Don't confuse them with the **elbow** and **knee** (see Fig. 1-1). Elongation of the foot is an adaptation for running because it lengthens the step and stride. The various segments of the limb can be determined by careful palpation. They correspond to our own: **brachium** (upper arm), **antebrachium** (forearm), and **hand** in the pectoral appendage; **thigh, crus** (shin), and **foot** in the pelvic appendage.

Unless your specimen is unusually old, body hair will not be conspicuous, but some hair can be found on the eyebrows, on the snout, and under the chin. You may notice a peeling layer of skin covering much of the body; this is the **epitrichum,** a layer of embryonic skin that is sloughed off as the hairs develop beneath it.

B. SKIN

Study the structure of the skin and hair on a microscopic slide of a vertical section through the skin of a mammal. The following account pertains to cat skin, which is the slide often available from biological supply companies, but it is applicable to that of many other mammals.

The skin consists of two layers: an outer, thin one, the **epidermis,** and an inner, much thicker one, the **dermis** (Fig. 1-3). Under high power, it can be seen that the epidermis is a stratified squamous epithelium. Its **basal layer,** next to the dermis, consists of elongate, columnar cells oriented more or less perpendicular to the plane of the skin surface. (Cell boundaries are often difficult to see in animal tissue, but the size and orientation of a cell can often be inferred from its more conspicuous nucleus.) New cells are produced in this basal layer; hence mitotic figures can sometimes be

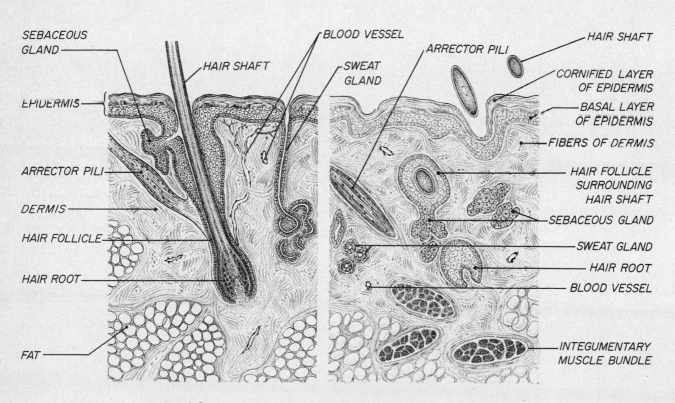

FIGURE 1-3
Microscopic vertical section through cat skin: *left*, diagrammatic section; *right*, typical appearance of a slide.

found in it. Many of the new cells move toward the surface, where they become flattened and filled with a protein known as keratin that is insoluble in water. Eventually they die and form a superficial horny, or **cornified, layer,** which gradually wears away. Well-defined intermediate layers can be recognized in some skin regions.

The dermis consists of a dense, fibrous connective tissue. Bundles of intercellular fibers can be seen running in many directions. Most of these fibers are composed of the protein collagen, which is relatively inelastic, but, some are composed of a different, more elastic protein. The cellular components of fibrous connective tissue are scattered **fibroblasts,** of which only the nuclei can be seen. Bundles of striated muscle fibers, representing the insertion of integumentary muscles that move the skin, are found in the deeper parts of the dermis in many mammals. Fat may also be found in this region and in the subcutaneous tissue.

Hair follicles extend into the dermis from the epidermis. (Because the plane of the section of a slide seldom parallels the axis of the hair, as it does in the left half of Figure 1-3, most of the hairs will be seen in cross or oblique sections, as in the right half of the figure.) A hair follicle consists of a stratified epithelium supported by connective tissue and continuous with the epidermis. Each follicle completely surrounds a **hair shaft**—a shaft of cornified epithelial cells that are continually produced at the base of the follicle. In some slides, bundles of smooth muscle fibers, **arrector pili,** can be seen. They are attached to the follicles and extend toward the skin surface. They pull the follicles, which are normally somewhat oblique to the plane of the skin surface, into a more nearly erect position, which enables the hairs to form a more efficient insulating layer when body temperature falls.

Two types of skin glands of epithelial origin are present. **Sebaceous glands** are associated with hair follicles in some regions of the body. If present in your slide, they appear as sac-like clumps of lighter-staining cells attached to, or lying beside, the follicles. Many of their cells are filled with oil droplets, which are discharged into the follicle when the cells break down. **Sweat glands** are long, coiled, tubular glands. They are not common in most of the skin of heavily furred mammals, but most slides include some. They appear in section as small clumps of cuboidal epithelial cells. The nuclei of the cells are relatively large and round. In examining a cross section, you may see that the cells surround a very small lumen, but shrinkage of the tissue in the course of slide preparation sometimes obliterates the cavity. Flattened nuclei peripheral to the cuboidal cells belong to elongated **myoepithelial cells,** whose contraction assists in the discharge of sweat. Sweat contains some salts and excretory products, but its main component, water, is secreted in abundance when body temperatures rise, and its evaporation removes much body heat. Many heavily furred mammals have few sweat glands and reduce body temperature by panting.

Other small, tubular structures in the skin are **blood vessels.** Blood vessels differ in appearance from sweat glands in having a relatively large lumen that is lined by a single layer of thin, squamous epithelial cells, whose nuclei are much smaller than those of cuboidal cells and are frequently flattened. The wall of a capillary consists only of a layer of squamous epithelial cells, but that of a small artery or vein will also contain some smooth muscle and connective tissue. The amount of blood flow through vessels near the surface of the dermis plays an important role in regulating body temperature. Body heat is conserved by a constriction of these vessels and a reduction of flow when ambient temperatures are low; opposite changes occur when temperatures are high.

C. SKELETON

Study the mammalian skeleton: a mounted skeleton of a human being and such quadruped skeletons as are available. The skeletons of all mammals are basically alike, as can be seen from the drawings of human and pig skeletons in Figures 1-4 and 1-7. They differ primarily with respect to meeting different stresses resulting from different methods of support and locomotion, for the bones serve primarily to resist compression stresses and to transmit the action of muscles by acting as levers. In addition, some bones—such as many bones of the skull, the vertebrae, and the ribs—house and protect delicate internal organs.

The skeleton of all vertebrates is internal. It comprises the **visceral skeleton,** associated in primitive vertebrates with the wall of the cranial part of the gut tube (i.e., the pharynx), and the **somatic skeleton,** associated with the body wall and appendages. The visceral skeleton is conspicuous in fishes, where it helps to form the gill apparatus and jaws, but it has become very much reduced in the evolution of terrestrial vertebrates. Its most conspicuous parts in mammals are the **hyoid bone,** which supports the base of the tongue, three **auditory ossicles** in the tympanic cavity (Exercise 6), and the cartilages of the larynx (Exercise 3). The somatic skeleton consists of the **axial skeleton,** in the longitudinal axis of the body (skull, vertebral column, ribs, and sternum), and the **appendicular skeleton** (bones of appendages and supporting girdles).

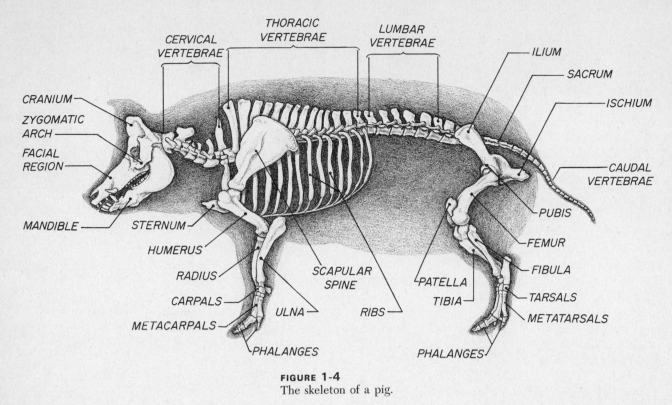

CERVICAL
VERTEBRAE

THORACIC
VERTEBRAE

LUMBAR
VERTEBRAE

ILIUM

SACRUM

ISCHIUM

CRANIUM

ZYGOMATIC
ARCH

FACIAL
REGION

CAUDAL
VERTEBRAE

MANDIBLE

STERNUM

HUMERUS

RADIUS

CARPALS

METACARPALS

PHALANGES

ULNA

SCAPULAR
SPINE

RIBS

PATELLA

TIBIA

PHALANGES

PUBIS

FEMUR

FIBULA

TARSALS

METATARSALS

FIGURE 1-4
The skeleton of a pig.

1. Skull

The skull consists of a **cranial region,** housing the brain and the inner and middle ear, and a **facial region,** containing the eyes and nose and forming the upper and lower **jaws.** The size of the cranial region is correlated with the size of the brain.

Observe the pair of large **orbits** that contain the eyes (Fig. 1-5A). A depression, the **temporal fossa,** for temporal jaw muslces lies posterior to each orbit. It is bounded inferiorly by a bar of bone, the **zygomatic arch,** that extends from the base of each orbit to the cranium. The large foramen just posterior to this arch is the **external acoustic meatus.** It leads into the **tympanic cavity,** or middle ear, but in life a tympanic membrane lies at the base of the meatus. The large bump of bone posterior to the external acoustic meatus is the **mastoid process** to which certain neck muscles attach. In life, a ligament extends from the pointed **styloid process,** located under the tympanic cavity, to the hyoid bone, which it helps to hold in place.

External nostrils lead from the front of the face into a pair of **nasal cavities.** The passage from the anteromedial corner of each orbit into the nasal cavity is the **nasolacrimal canal** which carries a duct through which tears are drained.

Examine the underside of the skull and notice that the nasal cavities are separated from the mouth by the **hard palate** (Fig. 1-5B). A pair of large **internal nostrils** (*choanae*) connect the nasal cavities with a part of the pharynx. The large opening through which the spinal cord enters the skull is the **foramen magnum.** It is flanked by a pair of rounded **occipital condyles** by which the skull articulates with the vertebral column. Most of the other foramina, all of them smaller, are for the passage of cranial nerves and blood vessels. The depression near the posterior end of the zygomatic arch, the **mandibular fossa,** is part of the jaw joint. The relatively large **carotid foramen** medial to the mandibular fossa is for the passage of the internal carotid artery, one of the main vessels supplying the brain. The small foramen between it and the mandibular fossa carries the **auditory tube** from the tympanic cavity to the pharynx. Verify this by probing.

Remove the top of the skull and look into the **cranial cavity,** whose shape closely follows that of the brain. The pair of large lateral butresses in its floor are the **otic capsules,** and each houses a delicate internal ear. Many of the the individual bones that constitute the skull are shown in Figure 1-5.

A temporal muscle (Exercise 2) extends from the temporal fossa to the **coronoid process** on the lower jaw, or **mandible.** This is the dorsalmost process. The masseter muscle extends from the zygomatic arch as

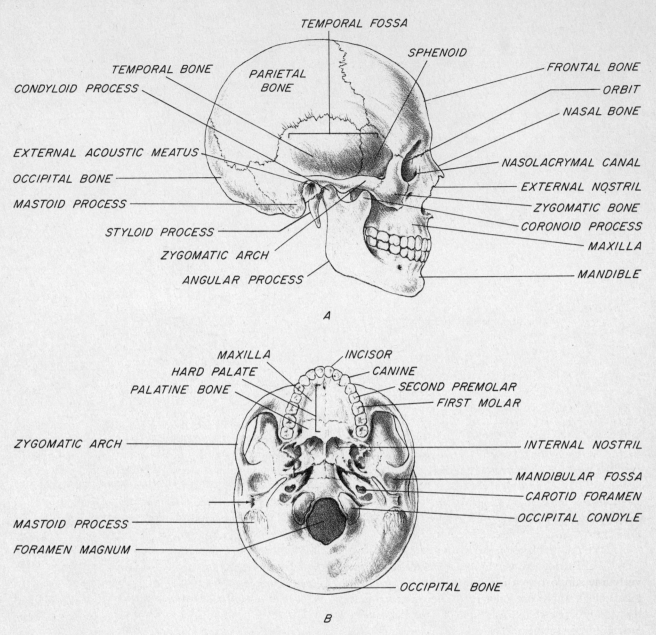

FIGURE 1-5

A human skull: (A) lateral view; (B) inferior view (the arrow enters the middle ear cavity through the external acoustic meatus and emerges from the auditory canal).

far as the ventralmost process, the **angular process.** By closing your jaws tightly you can feel these muscles bulging. The lower jaw articulates by its **condyloid process** with the mandibular fossa.

The teeth are deeply set in sockets in the jaw margins. On each side of each jaw of an adult human being, there are 2 **incisor teeth,** 1 **canine,** 2 **premolars** (bicuspids), and 3 **molars.**

2. Vertebrae, Ribs, and Sternum

Five vertebral regions can be recognized. Nearly all mammals have 7 **cervical vertebrae** in the neck; the first two of these, the **atlas** and the **axis,** are highly specialized to permit free movement of the head. Dorsoventral movements occur between the atlas and the occipital condyles; rotary ones, between the atlas

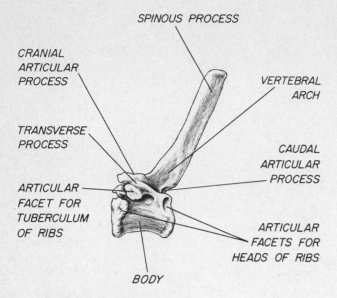

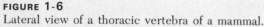

FIGURE 1-6
Lateral view of a thoracic vertebra of a mammal.

the vertebral column in correlation with the specific stresses to which each is subjected. **Intervertebral foramina,** through which spinal nerves pass, lie laterally between the vertebral arches. If an intervertebral disc slips out of place, the bodies of the adjacent vertebrae come closer together, the intervertebral foramen becomes smaller, and the nerve may be pinched.

Human beings have 12 pair of **ribs** (Fig. 1-7). All but the last two articulate by flexible **costal cartilages** to

and axis. Caudal to the cervical vertebrae are the **thoracic vertebrae,** to which ribs attach, and the **lumbar vertebrae,** without ribs. Their number varies: human beings have 12 thoracic and 5 lumbar vertebrae; the pig has 14 or 15 thoracic and 6 or 7 lumbar vertebrae. Caudal to the lumbar vertebrae are the **sacral vertebrae,** which are fused together to form a firm **sacrum** for the attachment of the pelvic girdle. Most mammals have 3 or 4 sacral vertebrae, but human beings have 5, the greater number being correlated with the greater problems of support inherent in a bipedal gait. **Caudal vertebrae** supporting a tail follow the sacrum. In human beings, there are only from 3 to 5 small vertebrae fused together to form a **coccyx** to which certain anal muscles attach.

A typical thoracic or lumbar vertebra consists of a large ventral disc, the **body,** dorsal to which is a neural, or **vertebral, arch** encasing the spinal cord (Figs. 1-6 and 7-1). **Intervertebral discs** separate successive vertebral bodies. The vertebral arches of successive vertebrae partly overlap and movably unite with one another by articular facets borne on **articular processes.** A pair of lateral **transverse processes** and a dorsal **spinous process** serve for muscular attachment. In addition, each transverse process of the thoracic vertebrae bears a facet that articulates with the tuberculum of a rib; other facets for the heads of ribs are on the body. The size of the vertebral bodies and the length and inclination of the spinous processes vary greatly along

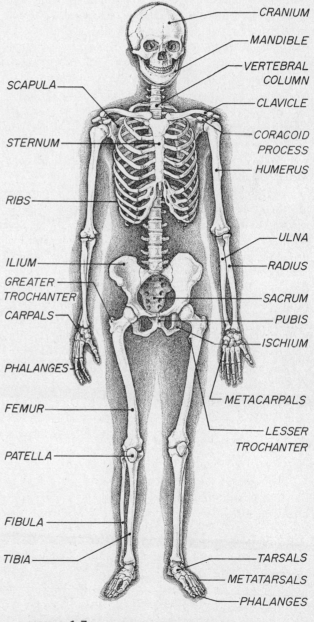

FIGURE 1-7
Anterior view of the skeleton of a human being.

the breast bone, or **sternum.** The thoracic basket thus formed protects the heart and lungs, and its movements play an important role in ventilating the lungs.

3. Appendicular Skeleton

The major element of the pectoral girdle is a large, blade-shaped **scapula** (Figs. 1-4 and 1-7). Forces are transmitted between it and the axial skeleton by a muscular sling; there is no bony attachment. It bears a long ridge on its lateral surface, the **scapular spine,** to which muscles attach, and a socket, the **glenoid cavity,** that receives the humerus of the upper arm. Cranial and median to the glenoid cavity is a beak-shaped process, the **coracoid process,** which is a remnant of a much larger bone in the girdle of lower terrestrial vertebrates. Human beings have a well-developed collarbone, or **clavicle,** extending from the scapula to the sternum, but this is missing or reduced in many quadrupeds. Its loss is an adaptation for running. Absence of the clavicle prevents the shoulder joint from being deflected as the trunk moves forward past a front limb that has been placed on the ground.

The bones of the pectoral appendage are a **humerus** in the upper arm, a **radius** and **ulna** in the forearm, **carpals** in the wrist, **metacarpals** in the palm, and **phalanges** supporting the free parts of the digits. The distal end of the radius is adjacent to the first finger: that is, it lies on the median side of the hand when the hand is in the normal quadruped position. The radius rotates on the ulna as the hand changes from a prone to a supine position in human beings, but this capacity is lost in the pig and other hoofed animals whose limbs simply move back and forth.

Each side of the pelvic girdle consists of three bones fused together. All meet in the socket, **acetabulum,** in which the leg articulates. An **ilium** extends dorsally and attaches to the sacrum. A **pubis** lies ventral and cranial to the acetabulum, and an **ischium** ventral and caudal to the acetabulum. In adult mammals all are fused together to form the hip bone, **os coxae.**

The bones of the pelvic appendage are a **femur** in the thigh, a stout **tibia** and thin **fibula** in the shank, **tarsals** in the ankle, **metatarsals** in the sole, and **phalanges** in the free part of the digits.

If a skeleton of a young mammal is available, notice the conspicuous transverse cleft near each end of the long limb bones. In life, a cartilaginous **epiphyseal plate** is located here. Its growth and replacement by bone is responsible for the growth in length of the bones.

There is much adaptive variation in both pectoral and pelvic appendages among different mammals according to their methods of locomotion, but the basic pattern described can be detected in all of them.

4. Bone

Although some cartilage is present, the major skeletal material of most adult vertebrates is bone. Bone lacks the elasticity of cartilage, but its resistance to compression is six times as strong as that of cartilage and to tension stresses is nine times as strong.

Living bones are covered by a layer of connective tissue called the **periosteum** (Fig. 1-8). **Compact bone** forms the periphery of a bone and **spongy bone** lies deep to this near the ends of the bone. Marrow fills the large cavities of bone. The bone itself is a type of connective tissue in which the intercellular matrix consists

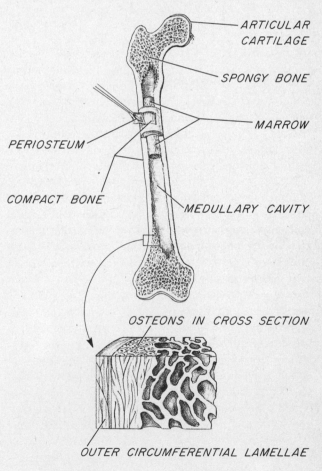

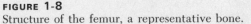

FIGURE 1-8
Structure of the femur, a representative bone.

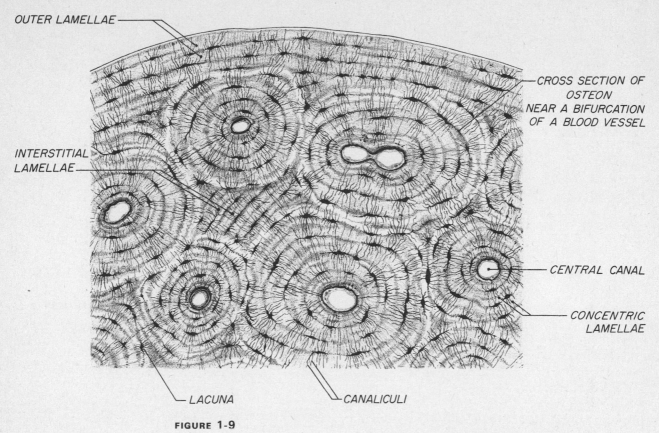

OUTER LAMELLAE

CROSS SECTION OF OSTEON NEAR A BIFURCATION OF A BLOOD VESSEL

INTERSTITIAL LAMELLAE

CENTRAL CANAL

CONCENTRIC LAMELLAE

LACUNA

CANALICULI

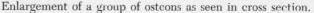

FIGURE 1-9
Enlargement of a group of osteons as seen in cross section.

of about 70 percent hard inorganic salts (mostly crystals of a calcium and phosphorous compound known as hydroxyapitite) and 30 percent fibrous proteins, chiefly collagen.

Examine a slide of a cross section of compact bone (Fig. 1-9). Most of the bony material is deposited in concentric rings around tiny blood vessels that tend to parallel the longitudinal axis of the bone. Each of these columnar units, which appear circular in cross section, is an **osteon,** or Haversian system. In the center of each is a **central canal** containing, in life, one or more blood

vessels (usually capillaries and venules) embedded in loose connective tissue. Around this canal are concentric layers of bone matrix, the **concentric lamellae.** Between the lamellae are rows of dark dots, the **lacunae,** which in life contain the cellular elements of the bone (**osteocytes**). Minute **canaliculi** extend more or less radially from the lacunae and contain processes of the osteocytes. Bone is a dynamic tissue and much of it is reabsorbed and rebuilt in the course of its development. Parts of former osteons (**interstitial lamellae**) can be seen between the ones now complete.

Muscles

Muscles constitute the largest of the organ systems. Besides bracing the bones and maintaining posture, they function in locomotion and in propelling materials through the digestive tract, blood vessels, and other passages of the body. Those which are confined to the walls of the gut and other visceral organs are called **visceral muscles.** Most of them are smooth muscles histologically and will not be studied, but visceral muscles in the anterior part of the digestive tract are more conspicuous, for they attach onto certain skeletal elements—the gill arches of fishes and the jaws and certain parts of the shoulder in terrestrial vertebrates. These anterior visceral muscles are known as **branchiomeric muscles.** Other muscles, which are located in the body wall and appendages, are called **somatic muscles.** Histologically, branchiomeric and somatic muscles are striated.

The muscles of the pig are quite representative of those of mammals. They should be studied by referring to as mature a specimen as possible and should be dissected with care for fetal muscles are softer than those of an adult and tear easily.

A. SKINNING

To remove the skin from the left side of the body, make a middorsal incision through the skin that extends from the back of the head to the base of the tail. While you are making the incision, pull up on the cut edge with forceps to avoid cutting underlying muscles. Make additional incisions that extend from the middorsal line ventral to the left ear and across the cheek to the chin, around the tail and genital area, and down the lateral surfaces of the left legs. Carefully peel off the skin from the body, leaving skin on the top of the head, feet, tail, and genital area, including the scrotum and penis if you have a male specimen. The loose connective tissue beneath the skin is known as **superficial fascia.** It can easily be torn with forceps. Do not injure deeper muscular tissue, which looks a bit darker and consists of parallel strands of fibers. A tough **deep fascia** encases some muscles and should be left on for the time being. As you remove the skin, cut the blood vessels and nerves that supply it. Those along the trunk emerge from the body at segmental intervals. This pattern is an indication of the basic segmentation of the vertebrate body, a feature less obvious in mammals than in fishes.

Look for muscle fibers posterior to the armpit that extend onto the lateral flank skin that you are removing. These fibers are parts of the **cutaneous trunci,** an extensive but inconspicuous sheet of muscle that fans out from its origin on chest muscles and the midventral line to insert into the trunk skin, which it can move. The muscle is present in most mammals but absent in human beings. Cut through its origin and remove it with the skin, but do not cut through a chest muscle extending toward the midventral line. The cutaneous trunci is a somatic muscle.

The ventral and lateral surfaces of the neck are covered by another integumentary muscle, the **platysma,** which continues onto the face where it breaks up into numerous **facial muscles of expression** associated with the eyelids, lips, nose, and auricles of the ears. Platysma and facial muscles are branchiomeric in origin, having evolved from gill arch muscles. Certain of these muscles make the cheeks and lips of mammals fleshy and give mammals their unique ability to suckle. Leave the platysma and facial muscle on the body until the skin is removed and then take them off very carefully. A thin, but extensive, salivary gland, the **parotid gland** (Fig. 3-1), lies on the side of the neck deep to the platysma. It may be studied now (see Exercise 3) or removed with the platysma and examined later on the other side of the body. The platysma and parotid gland lie superficial to a large vein (the **external jugular**) and its tributaries. Do not destroy them.

B. MUSCLE TERMINOLOGY

An individual muscle consists of a great many muscle fibers, or cells, that are held together by their investing connective tissue and have common attachments to a body part, usually a skeletal element. The attachment is made by the investing connective tissue, which is often conspicuous enough to form a cordlike **tendon** or a sheetlike **aponeurosis.** One end of a muscle, its **origin,** attaches to a skeletal element that is held in a fixed position when the muscle contracts; the other end, its **insertion,** attaches to a part that is free to move (Fig. 2-1). For a limb muscle, it is conventional to consider its proximal end as its origin and its distal end as the insertion. Origin and insertion are convenient descriptive terms, but their importance should not be overemphasized. The tension developed by a muscle is the same at each end, and in some cases whether one end or the other is the fixed end depends on the action being performed.

Because muscles perform work only by contracting, they are organized into antagonistic groups. One group moves a part in one direction and the antagonistic group, in the opposite direction. The following terms define the more common antagonistic actions:

(1) **Flexion** and **extension:** flexion is a bending; that is, it is the approximation of a distal part toward a more proximal part, such as that which occurs during the movement of the forearm toward the upper arm. Extension is the opposite action.

(2) **Protraction** and **retraction:** protraction is the forward movement of an entire limb at the shoulder or hip; although the term flexion is sometimes used for this movement, protraction is more accurate for a quadruped. Retraction is a backward movement of an entire limb.

(3) **Adduction** and **abduction:** adduction is the pulling of a part toward a point of reference; abduction is the movement away from the point of reference. The midventral line of the body is the point of reference for adduction and abduction of a limb.

Many students are overwhelmed by the names for muscles, but these names can be very helpful because they are generally descriptive of a muscle's attachments (sternomastoid), shape (trapezius), number of divisions (triceps), or function (tensor fasciae latae).

C. MUSCLE DISSECTION

Before attempting to identify muscles in any region, carefully clean up the area by picking away loose connective tissue and developing fat, looking for the separations between muscles. Separations can be recognized by a change in fiber direction, for the fibers of one muscle have common attachments different from those of adjacent muscles. As you find separations

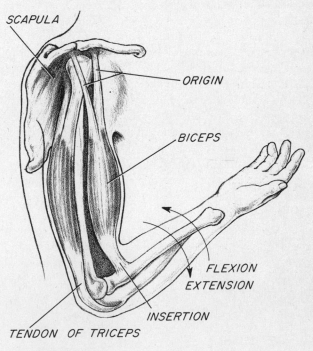

SCAPULA

ORIGIN

BICEPS

FLEXION
EXTENSION

INSERTION

TENDON OF TRICEPS

FIGURE 2-1
The attachments and actions of a representative set of antagonistic muscles in the human arm.

between muscles, pick away at them and make them clearer. Separate muscles from one another as completely as possible, but do not cut through any muscles until directed to do so. When you cut through a superficial muscle to reveal a deeper one, cut across its center, or **belly,** at right angles to its fiber direction, and turn back (reflect) the two ends. This procedure will facilitate reconstructing superficial muscles when you wish to review them.

D. MUSCLES OF THE SHOULDER

The first muscles to be considered are a group that extends from the trunk to the pectoral girdle and from the trunk and girdle to the proximal part of the humerus. Collectively, they move the shoulder and the arm as a whole, or, when the front foot remains in a fixed position on the ground, they move the trunk relative to the foot.

1. Superficial Muscles of the Shoulder

Clean up the lateral surface of the shoulder, neck, and upper arm and the ventral surface of the chest as directed above before attempting to identify these muscles. Reference to Figures 2-2 and 2-3 will help you find the separations between the muscles.

A thin, triangular-shaped **trapezius** arises from the dorsal surface of the neck and thorax and its fibers converge to insert on the scapular spine. The muscle helps to hold the scapula in place and to move it forward and backward. Most muscles of the shoulder are somatic, but the trapezius is branchiomeric, because it has evolved from fish gill arch muscles that acquired a secondary attachment onto the pectoral girdle.

The nearly band-shaped muscle cranial to the trapezius and extending from the back of the neck and head to its insertion on the proximal part of the humerus is the **brachiocephalicus.** It helps to protract the arm. In those mammals in which a clavicle is present, the

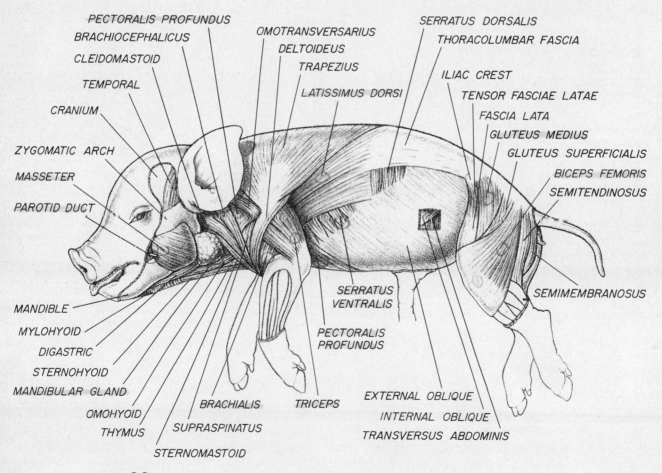

FIGURE 2-2
Lateral view of muscles of a pig after removal of integumentary muscles, with window cut in abdominal wall to show deeper muscle layers.

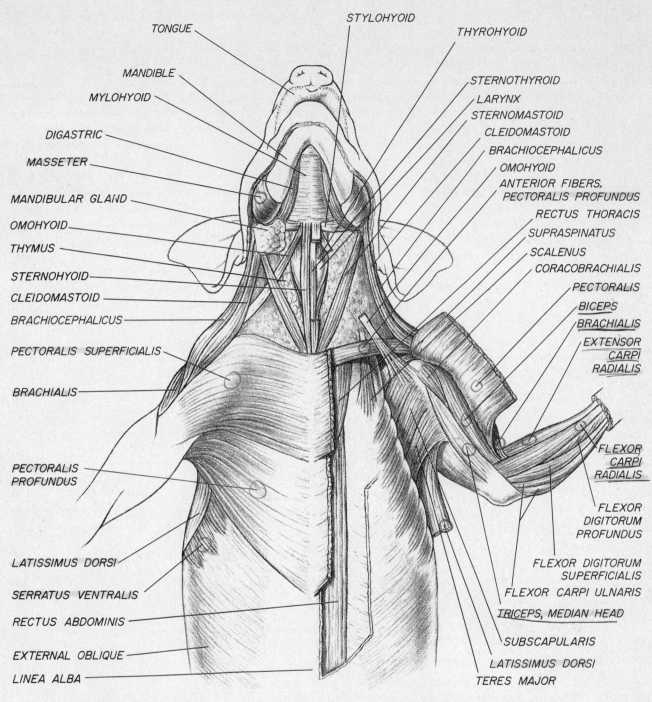

FIGURE 2-3
Ventral view of anterior muscles. Deeper muscles are shown on the right side of the drawing.

brachiocephalicus can be seen to be a compound of two muscles. The upper part, extending from the neck to the clavicle, is part of the trapezius complex (**cleidocervicalis,** or clavotrapezius); the lower, extending from the clavicle to the humerus, is part of the deltoid group (**cleidobrachialis,** or clavodeltoid). Cleidocer-

vicalis and cleidobrachialis join together in mammals that lose the clavicle. Although human beings have a clavicle, they do not have these divisions of the trapezius and deltoid.

A narrow, band-shaped **omotransversarius** can be found between the brachiocephalicus and trapezius.

It arises on the skull and atlas, inserts on the scapular spine, and helps to protract the scapula. Human beings do not have this muscle.

The **deltoideus** arises near the insertion of the trapezius from the scapular spine and the fascia over deeper muscles. It extends ventrally to insert on the humerus next to the insertion of the brachiocephalicus. It helps to protract the upper arm.

The large, triangular muscle that lies caudal to the brachium and fans out over the back is the <u>latissimus dorsi</u>. It arises from the dorsal surface of the thorax deep to the caudal part of the trapezius and from a tough sheet of deep fascia on the back (the **thoracolumbar fascia**). It passes into the armpit to insert on the proximal part of the humerus, deep to a large arm muscle, the triceps, which is described in Section E. The latissimus dorsi is a major retractor of the arm.

The ventral surface of the chest (Fig. 2-3) is covered by the large, triangular-shaped **pectoralis.** The more superficial and smaller <u>pectoralis superficialis</u> (which equals the pectoralis major of human anatomy) arises from the cranial half of the sternum and inserts along almost the entire length of the humerus and on the fascia covering the proximal part of the forearm. It adducts and retracts the arm. The deeper and larger <u>pectoralis profundus</u> (which equals the pectoralis minor of human beings) arises from the entire length of the sternum. Its fibers extend craniolaterally deep to those of the pectoralis superficialis. Trace them by cutting through the superficial pectoral. The caudal fibers insert on the proximal part of the humerus; the cranial ones extend in front of the shoulder joint and curve over the lateral edge of the scapula to insert on fascia covering a deep shoulder muscle (the **supraspinatus**) as shown in Figures 2-2 and 2-4.

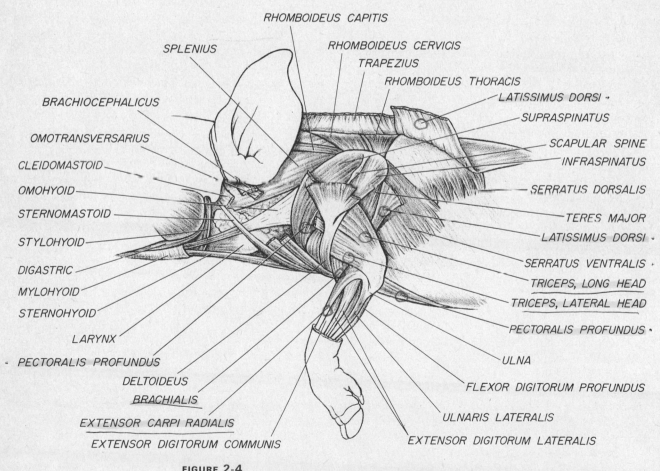

FIGURE 2-4
Lateral view of deeper muscles of the neck and shoulder.

2. Deep Muscles of the Shoulder

Cut through and reflect the trapezius, omotransversarius, and latissimus dorsi (Fig. 2-4). Find the scapular insertion of the cranial fibers of the pectoralis profundus. Cut through it to expose the thick **supraspinatus** that arises from the lateral surface of the scapula cranial to the scapular spine. An **infraspinatus** arises from the lateral surface of the scapula caudal to the spine. You will have to reflect some of the deltoideus to see it clearly. Both muscles insert on the proximal end of the humerus. The supraspinatus helps to protract the arm; the infraspinatus rotates it outward.

Trace the latissimus dorsi as it passes medial to the triceps to its humeral insertion. The muscle fibers arising from the caudal border of the scapula and inserting with the latissimus dorsi constitute the **teres major.** This muscle helps to retract the arm. A small **teres minor,** which is difficult to find, arises from the border of the scapula between the infraspinatus and the lateral surface of the triceps. It may be omitted.

Three muscles extend from the dorsal, or vertebral, border of the scapula to their origin on the back of the skull and the spinous processes of the cervical and anterior thoracic vertebrae. They are, from cranial to caudal, the **rhomboideus capitis, rhomboideus cervicis,** and **rhomboideus thoracis.** All act on the scapula, which they help to hold in place and to pull cranially or caudally. Human beings lack the rhomboideus capitis.

Cut through the rhomboids and pull the dorsal border of the scapula laterally. The serratus ventralis is the large, fan-shaped muscle arising by a series of slips from the ribs and from the transverse processes of the cervical vertebrae. It attaches on the dorsal border of the scapula deep to the insertion of the rhomboids. In addition to moving the scapula backward and forward, the serratus ventralis acts as a muscular sling to transfer body weight from the trunk to the pectoral girdle and appendage. Recall that the pectoral girdle does not articulate with the vertebral column.

Turn the specimen on its back and cut through the pectoralis profundus, except for its cranial fibers extending to the scapula (Fig. 2-3). Clean up the medial surface of the scapula. A neck muscle, the **omohyoid** (Section F), attaches onto the fascia on the medial surface of the scapula. Observe this and then cut through it. Do not injure blood vessels and a group of whitish strands (nerves of the brachial plexus) going to the arm. Find the serratus ventralis, latissimus dorsi, teres major, and the edge of the supraspinatus in this view. The thick muscle arising from the medial surface of the scapula between the supraspinatus and teres major is the **subscapularis.** It inserts upon the proximal end of

the humerus and acts primarily as an adductor of the arm.

E. MUSCLES OF THE ARM

The following muscles cover the humerus. Most arise from the humerus or girdle and act on the forearm. Largest is the **triceps brachii,** whose location has been seen. The medial surface of the triceps is covered by a thin **tensor fasciae antebrachii** that arises from the edge of the latissimus dorsi and inserts on the antebrachial fascia and on the insertion of the triceps. Cut through it if it has not already been destroyed. The triceps brachii arises by three heads—a long head from the caudal border of the scapula (Fig. 2-4), a lateral head from the proximal part of the lateral surface of the humerus, and a medial head (Fig. 2-3) from the proximal half of the medial surface of the humerus. All converge to form a large, common tendon that passes over the elbow to insert on the proximal end of the ulna. The triceps is the extensor of the forearm.

Two smaller muscles lie on the cranioventral surface of the humerus, laterally a **brachialis** and medially a **biceps brachii** (Figs. 2-3 and 2-4). The insertions of the brachiocephalicus and pectoralis pass between them. The brachialis arises from the humerus and inserts upon the ulna. The biceps brachii, which is two-headed in human beings, arises in the pig by a single tendon that passes through a groove on the craniomedial surface of the humerus to attach onto a small process (**coracoid process**) on the scapula. This process lies just medial and cranial to the glenoid cavity. The biceps inserts on the radius and ulna. Both brachialis and biceps are flexors of the forearm.

A small **coracobrachialis** (Fig. 2-3) arises from the coracoid process, crosses the medial surface of the shoulder joint, and inserts on the humeral shaft. It is a weak adductor of the arm, and it also reinforces the shoulder joint.

Forearm muscles are not described, but many can be identified from Figures 2-3 and 2-4.

F. LATERAL NECK MUSCLES

Muscles on the lateral surface of the neck extend from the pectoral girdle and sternum to the skull. Their primary function is to flex and turn the head.

Again find the brachiocephalicus, which has been described with the shoulder muscles. The **cleidomastoid** is a band-shaped muscle that lies cranial to the brachiocephalicus (Fig. 2-2). It attaches to the deep surface of the brachiocephalicus at a point at which the clavicle would be if one were present and then extends cranially and dorsally to the mastoid process of the skull, which is located ventral to the ear. A narrower **sternomastoid** lies cranial to the cleidomastoid and extends between the front of the sternum and the mastoid process.

Carefully dissect between the sternomastoid and the midventral region of the neck (Figs. 2-3 and 2-4). A large thymus lies on either side of the rather firm voice box, or larynx. The thymus is crossed by a very thin, band-shaped **omohyoid,** which extends between the hyoid bone at the base of the tongue and the fascia covering the medial surface of the scapula. It passes deep to the sternomastoid, and its scapular attachment has already been seen and destroyed (Section D). The sternomastoid, cleidomastoid, and omohyoid have a branchiomeric origin, in common with the trapezius.

G. THROAT MUSCLES

Several ribbon-shaped muscles extend from the sternum to the hyoid and larynx on the ventral side of the neck (Fig. 2-3). They belong to a subgroup of somatic muscles called the **hypobranchial group** because they lie ventral to the gills in fishes. Those described here pull the larynx and hyoid bone caudally, an action that occurs in swallowing. The names of the muscles are descriptive of their origins and insertions. Most ventral of the group is a **sternohyoid.** Cut through and reflect the sternohyoid and omohyoid on one side of the body. A long **sternothyroid** lies deep to the sternohyoid and thymus and covers the thyroid gland and windpipe, or trachea. Its insertion is on the thyroid cartilage, one of the cartilages of the larynx. A shorter **thyrohyoid** extends from the insertion of the sternothyroid to the hyoid.

More cranial hypobranchial muscles, which will not be described, lie deep between the hyoid and chin and extend into the tongue. They pull the hyoid cranially, move the tongue, and help to open the jaw.

H. HEAD MUSCLES

Head muscles are branchiomeric in origin. Most function in jaw movements; a few function in swallowing.

Remove skin from the side of the skull between the eyelids and auricle (Fig. 2-2). Also remove a facial muscle extending from the skull to the anterior part of the auricle. A **temporal** muscle lies beneath it. It arises from the temporal fossa of the skull and passes deep to the zygomatic arch to insert on the coronoid process of the mandible. A large **masseter** arises from the ventral margin of the zygomatic arch and inserts on the lateral surface and the angular process of the mandible. Temporal and masseter are major jaw-closing muscles. Smaller pterygoid muscles arise deep from the base of the skull and go to the medial side of the jaw. They will not be seen in this dissection.

A **digastric** (Figs. 2-2 and 2-3) inserts along the medioventral margin of the mandible. Reflect the mandibular salivary gland located at the angle of the mandible and trace the digastric cranially and dorsally. It forms a strong tendon that arises from the base of the skull. The digastric is the major jaw-opening muscle. It derives its name from its condition in human beings where there are two bellies; the craniodorsal part is also fleshy and is separated from the cranioventral part by a short tendon.

The **mylohyoid** consists of a thin sheet of transverse fibers lying in the floor of the mouth between the digastric muscles of opposite sides of the body. It arises from the mandible deep to the insertion of the digastric and inserts on a midventral septum of connective tissue and on the hyoid. The mylohyoid compresses the floor of the mouth and assists in swallowing.

I. TRUNK MUSCLES

Trunk muscles are somatic. They can be divided into an **epaxial group,** which lies dorsal and lateral to the vertebral column and acts primarily to brace and move the back, and a **hypaxial group,** which contributes to the lateral and ventral parts of the trunk wall.

Three thin sheets of hypaxial muscle lie in the abdominal wall. They arise dorsally from the ribs and thoracolumbar fascia and insert ventrally by aponeuroses to a midventral connective tissue septum, the **linea alba.** They can be distinguished by their position in relation to one another and by their fiber direction (Fig. 2-2). Fibers of the superficial layer, the **external oblique,** extend caudally and ventrally. Cut a window in this layer on the upper part of the lateral surface of the abdomen, and notice that the fibers of the **internal oblique** extend cranially and ventrally nearly at right angles to those of the external oblique. Separate the fibers of the external oblique and note that the fibers of the **transversus abdominis** lie in nearly the transverse plane. All three layers help to support the abdominal wall and to compress the viscera during expiration.

A longitudinal band of muscle, the **rectus abdominis** (Fig. 2-3), lies deep to the external oblique and extends between the pelvic girdle and ribs. Its cranial portion is tendinous. It too helps to support the abdominal wall.

Thoracic hypaxial muscles pull the ribs either forward to enlarge the thoracic cavity during inspiration or backward to compress the cavity during expiration. The major inspiratory muscle, however, is the diaphragm, which will not be seen until the body cavity has been opened (Exercise 3). A short **rectus thoracis** (Fig. 2-3) extends from the sternum across the cranial end of the rectus abdominis to attach to cranial ribs, which it pulls caudally. A **scalenus** lies lateral and dorsal to the rectus thoracis. It arises from cervical vertebrae, crosses the cranial part of the insertion of the serratus ventralis, and attaches to several cranial ribs, which it pulls forward.

Pull the vertebral border of the scapula laterally and examine the chest wall median to the insertion of the serratus ventralis. The muscle fibers that extend from one rib caudally and ventrally to the next caudal rib constitute an **external intercostal.** Carefully cut through an external intercostal and you will find an **internal intercostal** with fibers at right angles to those of the external layer. External intercostals are inspiratory; internal intercostals are expiratory. These thoracic layers correspond to the external and internal oblique layers in the abdominal wall. There is only a trace of the transverse layer (**transversus thoracis**) in the thoracic wall. It lies near the midventral wall and may be seen when the thorax is opened (Fig. 3-5).

The thin sheet of muscle arising from the dorsal fascia and inserting on the dorsal part of the ribs, which it covers, is the **serratus dorsalis** (Fig. 2-4). It can be divided into cranial inspiratory and caudal expiratory parts.

If you cut through and reflect the serratus dorsalis and the thoracolumbar fascia, you will see the epaxial trunk muscles. Most form powerful, longitudinal bands that constitute the **erector spinae** group. A triangular **splenius** (Fig. 2-4) lies deep to the rhomboideus capitis and helps to support and turn the head.

J. MUSCLES OF THE PELVIS AND THIGH

Unlike the arrangement in the pectoral region, where all but a few muscles act across a single joint, many of the pelvic muscles extend across both hip and knee joints and therefore can move both the thigh and the shank. When the hind foot is held on the ground, these muscles move the trunk relative to the foot.

1. Lateral Muscles of the Hip and Thigh

Six muscles can be seen on the lateral surface of the hip and thigh (Fig. 2-2). A **tensor fasciae latae** arises

from the cranial end of the ilium and fans out to insert on the white sheet of fascia (**fascia lata**) near the knee. Because a part of this fascia, in turn, attaches onto the tibia, the tensor fasciae latae helps to extend the shank as well as to protract the thigh at the hip joint.

A very broad **biceps femoris** covers most of the lateral surface of the thigh caudal to the fascia lata. It arises from the caudal part of the sacrum and ischium and inserts by an aponeurosis along most of the length of the tibia. It forms the lateral wall of the **popliteal fossa,** the depression behind the knee joint. The biceps femoris acts across both the hip and knee joints, and so it retracts the thigh and flexes the shank.

Two gluteal muscles lie between the tensor fasciae latae and the biceps femoris. The **gluteus superficialis** is difficult to separate from the biceps femoris in the pig, for it too arises from the sacrum and its insertion fuses with the cranial margin of the biceps (Fig. 2-2). By careful dissection, the two can be separated at their origins, because part of the origin of the gluteus superficialis goes deep to that of the biceps. The superficial gluteal is quite distinct in human beings and forms the large buttock muscle (our gluteus maximus). The **gluteus medius** is more conspicuous in the pig. It arises from the lateral surface of the ilium. Its fibers pass caudoventrally, go deep to the gluteus superficialis and biceps, and insert on the greater trochanter of the femur dorsal to the hip joint. This insertion enables the gluteus medius to retract the thigh, as well as to act as a thigh abductor.

A band-shaped **semitendinosus** lies caudal to the biceps femoris. It arises from the ischium and inserts on the medial side of the tibial shaft where it helps to form the medial wall of the popliteal fossa. You can feel its powerful tendon of insertion on the caudomedial border of the fossa in your own leg. The semitendinosus assists the biceps femoris in thigh retraction and shank flexion. A **semimembranosus** lies caudal and medial to the semitendinosus, but it can be seen more clearly in a median dissection of the thigh (Fig. 2-6).

Cut through and reflect the tensor fasciae latae and biceps femoris (Fig. 2-5). The insertion of the gluteus medius can now be seen clearly. Carefully cut through the belly of the gluteus medius and you will find a deeper layer of the medius, which is sometimes called a **gluteus accessorius.** Its attachments and actions are similar to those of the medius. The piriformis, a distinct muscle in many mammals, appears to be incorporated in the gluteus medius of the pig.

Cut through and reflect the gluteus accessorius. A large **ischiatic nerve,** which supplies many of the hip and leg muscles, lies deep to the gluteus accessorius and superficial to a large, fan-shaped **gluteus profundus**

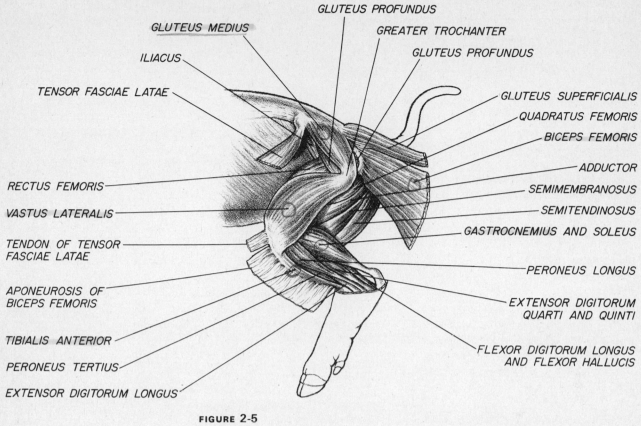

GLUTEUS PROFUNDUS
GLUTEUS MEDIUS
GREATER TROCHANTER
GLUTEUS PROFUNDUS
ILIACUS
TENSOR FASCIAE LATAE
GLUTEUS SUPERFICIALIS
QUADRATUS FEMORIS
BICEPS FEMORIS
ADDUCTOR
RECTUS FEMORIS
SEMIMEMBRANOSUS
VASTUS LATERALIS
SEMITENDINOSUS
GASTROCNEMIUS AND SOLEUS
TENDON OF TENSOR
FASCIAE LATAE
PERONEUS LONGUS
APONEUROSIS OF
BICEPS FEMORIS
EXTENSOR DIGITORUM
QUARTI AND QUINTI
TIBIALIS ANTERIOR
FLEXOR DIGITORUM LONGUS
AND FLEXOR HALLUCIS
PERONEUS TERTIUS
EXTENSOR DIGITORUM LONGUS

FIGURE 2-5
Lateral view of deeper pelvic and leg muscles.

(equivalent to our gluteus minimus). The gluteus profundus arises from the ilium and ischium cranial and dorsal to the acetabulum, and its fibers converge to insert on the greater trochanter slightly distal to the insertion of the gluteus medius. This muscle too is an adductor of the thigh.

Several small and deep muscles (**gemelli, obturators,** and **quadratus femoris**) arise from the periphery of the obturator foramen of the pelvis and from the membrane covering the foramen. All insert on the proximal end of the femur between the greater and lesser trochanters and on the lesser trochanter. They are difficult to separate, but part of the group can be seen in Figure 2-5. They rotate the thigh and assist in its retraction.

2. Medial Thigh Muscles

A thin, broad, muscular sheet, the **gracilis,** covers the caudomedial half of the thigh (Fig. 2-6). The gracilis arises from the pelvic symphysis and from the surface of a deeper thigh muscle (the adductor), and it inserts

by an aponeurosis along most of the length of the tibia. It is an adductor and retractor of the thigh and a flexor of the shank.

The **sartorius** is a narrow, thin band of muscle that partly covers the femoral blood vessels just cranial to the gracilis. It arises partly from iliac fascia and partly from the surface of a deep trunk muscle lying cranial to the pelvic girdle. (The latter part of the origin cannot be seen at this time.) It inserts on the proximal end of the tibia and helps to adduct the thigh and extend the shank.

Cut through and reflect the gracilis and sartorius. Most of the caudomedial part of the thigh is occupied by the large **semimembranosus** (Fig. 2-6), part of which has been seen previously from the lateral surface. The semimembranosus arises from the caudal part of the ischium and inserts on the distal end of the femur and adjacent parts of the tibia. Together with the semitendinosus, it forms the medial wall of the popliteal fossa. It acts primarily as a retractor of the thigh.

A triangular-shaped **adductor** lies cranial to the semimembranosus. It arises from the ventral surface

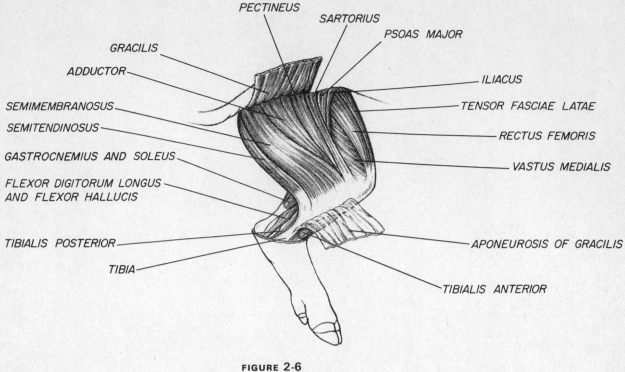

PECTINEUS
SARTORIUS
PSOAS MAJOR
GRACILIS
ADDUCTOR
SEMIMEMBRANOSUS
SEMITENDINOSUS
GASTROCNEMIUS AND SOLEUS
FLEXOR DIGITORUM LONGUS
AND FLEXOR HALLUCIS
TIBIALIS POSTERIOR
TIBIA
ILIACUS
TENSOR FASCIAE LATAE
RECTUS FEMORIS
VASTUS MEDIALIS
APONEUROSIS OF GRACILIS
TIBIALIS ANTERIOR

FIGURE 2-6
Medial view of leg muscles.

of the pubis and ischium and its fibers converge to insert along much of the length of the femoral shaft. It is not subdivided in the pig as it is in human beings and many other mammals. As its name implies, it is a femoral adductor, but it also assists in thigh retraction.

A smaller, triangular-shaped **pectineus** lies cranial to the adductor. It arises from the front of the pubis and inserts on the femur just cranial to the insertion of the adductor. It too adducts the thigh.

Two smaller muscles can be seen cranial to the origin of the pectineus. The more lateral **iliacus** arises from the ventral border of the ilium. The origin of the more medial **psoas major** from the ventral surfaces of lumbar vertebrae cannot be seen at this time. The fibers of both muscles extend caudolaterally, go deep to the pectineus, and insert in common on the lesser trochanter of the femur. These muscles help to protract and rotate the thigh.

3. Cranial Thigh Muscles

The cranial part of the thigh is occupied by a large **quadriceps femoris.** The four components of this complex converge to form a common **patellar tendon** that crosses the front of the knee joint to insert on the proximal end of the tibia. This tendon slides easily across the joint because it contains a small bone, the knee cap, or **patella.** The quadriceps femoris is the major

extensor of the shank. One of its heads, the **vastus lateralis,** can be seen on the lateral surface of the thigh (Fig. 2-5), in that it arises from the greater trochanter and much of the lateral surface of the femur. In a median view, the **vastus medialis** can be seen arising from the medial surface of the femur (Fig. 2-6). The **rectus femoris** lies between these vasti. Because it takes its origin from the pelvis just cranial to the acetabulum, it also helps to protract the thigh. The small fourth head, the **vastus intermedius,** arises from the craniolateral surface of the femur. It can be found by dissecting between the vastus lateralis and rectus femoris.

Shank muscles are not described, but many can be seen in Figures 2-5 and 2-6.

K. STRIATED MUSCLE

The skeletal muscles that you have been dissecting have a characteristic striped appearance when studied microscopically and so they are known as **striated muscle.** Other histological types are **cardiac muscle** of the heart wall, which is a special type of striated muscle, and **smooth muscle,** which is associated with the walls of blood vessels and visceral organs. Smooth muscle will be seen when a section of the intestine is examined (Exercise 3).

Examine a microscope slide of striated muscle. The individual muscle cells are very long with a uniform

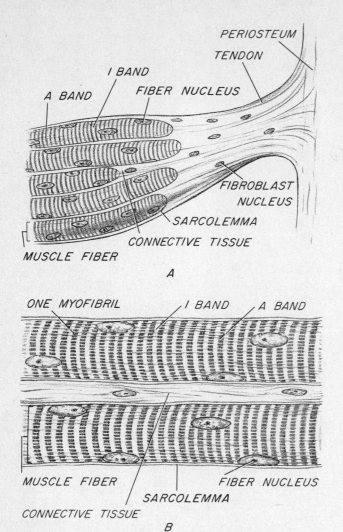

PERIOSTEUM
TENDON
I BAND
A BAND
FIBER NUCLEUS
FIBROBLAST NUCLEUS
SARCOLEMMA
CONNECTIVE TISSUE
MUSCLE FIBER

A

ONE MYOFIBRIL   I BAND   A BAND

MUSCLE FIBER   FIBER NUCLEUS
SARCOLEMMA
CONNECTIVE TISSUE

B

FIGURE 2-7
Microscopic lateral views of sections of striated muscle:
(A) a group of striated muscle fibers and the surrounding
connective tissue near the end of a muscle; (B) an
enlarged part of two adjacent fibers in which individual
myofibrils can be seen.

diameter ranging from 10 to 100 micrometers. Because of their great length, these cells are called **muscle fibers.** In a few cases the fibers may be as long as the muscle, but usually they are only a few centimeters long. Their blunt ends terminate in the connective tissue that permeates the muscle, surrounds individual fibers, and continues to the bones as tendons (Fig. 2-7A). Many elongated nuclei are present in a single fiber, usually just inside the cell membrane, or **sarcolemma.** The cytoplasm, called **sarcoplasm** in these fibers, is largely composed of longitudinal and banded **myofibrils.** If part of the muscle fiber has been torn during the preparation of the slide, you may see some of the individual myofibrils. They are barely visible with light microscopy because they are only 1 or 2 micrometers in diameter (Fig. 2-7B). The alternating dark and light bands that occur along a myofibril are known respectively as **A** and **I** bands because of their anisotropic and isotropic properties in polarized light. The A and I bands of one myofibril are in register with those of adjacent ones, and so the entire fiber has a banded or striated appearance.

Electron microscopic and biochemical studies have shown that the myofibrils, in turn, are composed of many ultramicroscopic **myofilaments** of two types. Thicker **myosin filaments** are limited to the dark A bands. Thinner **actin filaments** occupy the light I bands and extend a variable distance into the adjacent A bands. When a muscle fiber contracts, interactions between the actin and myosin filaments pull the actin filaments deeper into the A bands, and the I bands consequently become narrower.

Digestive and
Respiratory Systems

The digestive and respiratory systems are the organ systems through which food, water, oxygen, and all other essential raw materials enter the bodies of mammals, and the systems by which certain waste products of metabolism leave—carbon dioxide through the lungs and bile pigments with the feces. Most fecal materials, however, are roughage and bacteria, not by-products of cellular metabolism. The nitrogenous wastes of metabolism are removed by the excretory system (Exercise 5).

In all vertebrates, the respiratory system develops embryonically as an outgrowth of the pharyngeal part of the digestive tract; so it is convenient to study at least part of the structure of these two systems together.

A. THE HEAD AND NECK

1. Salivary Glands

If you did not see the salivary glands when dissecting the muscles, carefully remove the skin from one side of the head and neck over the area shown in Figure 3-1 in order to expose them, or study the glands in a demonstration dissection. If you remove the skin yourself, you must cut deeply enough to remove the facial muscles that attach to the skin in this region, but be careful not to cut into the salivary glands themselves. Muscle tissue can be recognized by the small, parallel bundles of muscle fibers that can be seen when the connective tissue is carefully picked off; glandular tissue has a different texture, consisting of little nodules of tissue clustered in bunches.

The **parotid gland** is a large, thin, approximately triangular gland occupying the area between the base of the ear, the shoulder, and the angle of the lower jaw. The angle of the lower jaw is covered in this region by a large jaw muscle, the **masseter.** The **parotid duct** extends forward from the cranioventral corner of the parotid gland around the lower border of the masseter and perforates the upper lip. Do not confuse the duct with branches of the facial nerve that emerge from beneath the cranial border of the parotid gland and supply the facial muscle. Small **buccal glands** lie beneath the skin over the lips. They are not large and are easily overlooked. Most of the **mandibular gland** lies beneath the parotid gland and just caudal to the angle of the jaw. Its duct passes medially to the angle of the jaw and can be followed forward by carefully removing the small jaw muscles beneath which it passes. It opens into the mouth ventral to the tongue at the same point as do minute ducts from the **sublingual gland.** The sublingual gland lies median to the lower jaw along the course of the mandibular duct.

The combined secretions of all these glands are the saliva, a complex solution containing an amylase (ptyalin) that initiates the breakdown of carbohydrates,

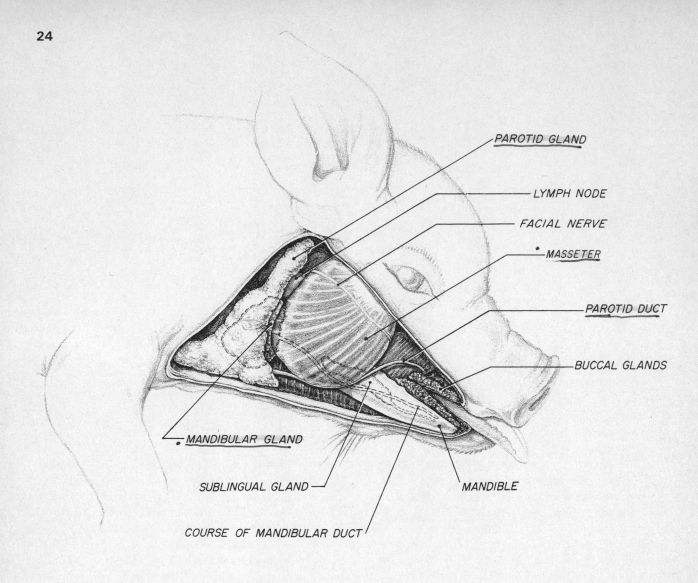

PAROTID GLAND

LYMPH NODE

FACIAL NERVE

MASSETER

PAROTID DUCT

BUCCAL GLANDS

MANDIBULAR GLAND

SUBLINGUAL GLAND

MANDIBLE

COURSE OF MANDIBULAR DUCT

FIGURE 3-1
Lateral view of salivary glands of the pig.

chloride ions necessary for the action of ptyalin, and water and mucus that helps lubricate the food and facilitates swallowing.

2. Mouth, Pharynx, Larynx, and Neck

Expose the organs of the mouth and pharynx by inserting a pair of scissors in the angle of the lips on the side not previously dissected and cutting posteriorly. If you have a large specimen, you may need to cut through the cheek on each side. Open the mouth as you do so, and make the incision follow the curvature of the tongue. Do not cut into the roof of the mouth. Continue the incision until you see a little flap of tissue, the epiglottis, extending dorsally to the free border of the soft palate. Carefully pull this down and continue

your incision dorsal to it and on into the gullet, or esophagus. The floor of the mouth and pharynx can now be swung open (Fig. 3-2).

Certain of the **teeth** may have emerged through the gums; others may form bulges beneath them. A mammal's teeth, in addition to helping the animal obtain food, play an important role in its mechanical breakdown. The **tongue** helps manipulate the food, pushing it between the teeth, mixing it with saliva, rolling it up into a ball, and pushing it back into the pharynx. **Papillae** of various shapes will be seen on its surface, particularly along its margins and at its base. Microscopic taste buds are associated with them.

The roof of the mouth consists of an anterior **hard palate** underlain by cartilage and bone and a more posterior, fleshy, **soft palate.** The **oral cavity** proper lies ventral to the hard palate.

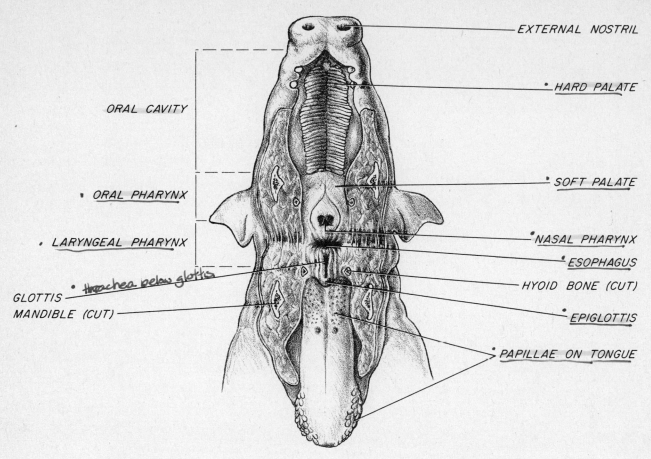

ORAL CAVITY

• ORAL PHARYNX

• LARYNGEAL PHARYNX

GLOTTIS
MANDIBLE (CUT)

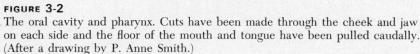
• trachea below glottis

EXTERNAL NOSTRIL

• HARD PALATE

• SOFT PALATE

• NASAL PHARYNX
• ESOPHAGUS
HYOID BONE (CUT)
• EPIGLOTTIS

• PAPILLAE ON TONGUE

FIGURE 3-2
The oral cavity and pharynx. Cuts have been made through the cheek and jaw
on each side and the floor of the mouth and tongue have been pulled caudally.
(After a drawing by P. Anne Smith.)

Paired **nasal cavities** lie dorsal to the hard palate and can be seen in a demonstration of a sagittal section of the head (Fig. 3-3). A vertical **nasal septum** separates the two cavities, and each cavity is largely filled with folds of tissue, the **conchae,** that increase the surface area available for olfaction and conditioning the inspired air. Further information on the nose is presented in Exercise 6.

The **pharynx** is the part of the digestive tract into which the gill slits open in a primitive vertebrate such as a fish. Certain gill pouches develop in this region in an embryonic mammal, but as the embryo develops they regress or are transformed into other structures. In an adult, the pharynx consists of an **oral pharynx** posterior to the oral cavity and ventral to the soft palate, a **nasal pharynx** dorsal to the soft palate, and a **laryngeal pharynx** caudal to the tongue, where the oral and nasal pharynges come together.

Make a longitudinal incision through the midline of

the soft palate to expose the nasal pharynx. Look for a small pair of slits, the orifices of the eustachian, or **auditory, tubes** in the dorsolateral walls of the nasal pharynx. They may be seen more clearly in a demonstration of a sagittal section of the head (Fig. 3-3). They develop from the first pair of gill pouches of the embryo, lead to the tympanic cavities, and help equalize air pressure on the ear drums. The nasal cavities open by a pair of **internal nostrils** (*choanae*) into the front of the nasal pharynx. The internal nostrils can be seen best on a skull.

The gullet, or **esophagus,** extends caudally from the laryngeal pharynx, and the **larynx** lies in the floor of this part of the pharynx. Carefully cut tissue away from the lateral and ventral surfaces of the larynx so that it is exposed on three of its sides. This will be facilitated by making a midventral incision down the neck.

In making this dissection, you will expose a large, glandlike mass on each side of the neck, the **thymus,**

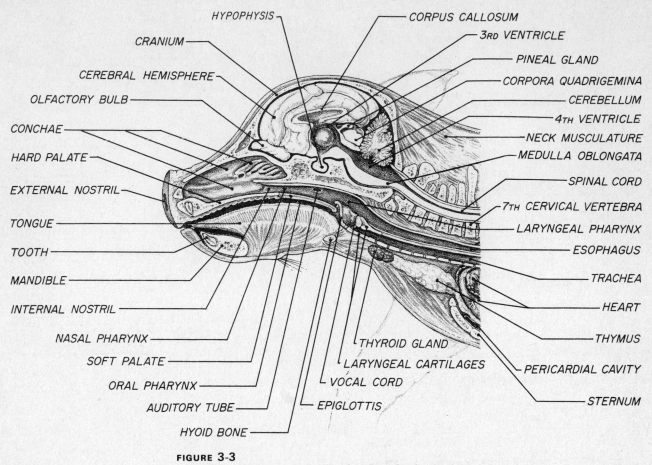

FIGURE 3-3
Sagittal section through the head and neck of a fetal pig.

which develops from certain gill pouches. It is relatively large in the fetus and infant, but regresses with age. It is important in the development of the body's defense mechanisms, because certain types of white blood cells (T-lymphocytes) mature here before they are distributed to lymph nodes and other lymphoid tissues. T-lymphocytes are responsible for certain immune reactions, particularly those associated with fungal and certain viral infections, and for rejecting foreign tissue. Their action must be suppressed in organ transplants.

As tissue is cleared from the larynx, notice that it is supported by several large cartilages that are part of the visceral skeleton. The **hyoid bone** (Exercise 1) may be seen cranial to it, embedded in the base of the tongue, and the windpipe, or **trachea,** extends caudally from it. The pharyngeal entrance to the larynx is protected by a trough-shaped flap of tissue, the **epiglottis.** The epiglottis helps to prevent food from entering the larynx by deflecting the food into the esophagus and moving back over the larynx entrance when the larynx

is pulled forward during swallowing. (You can demonstrate this movement of the larynx by putting your fingers on your own larynx, or Adam's apple, and swallowing.) Cut open the larynx by making a longitudinal, middorsal incision, and spread it open. The pair of small, whitish folds on its lateral walls are the **vocal cords.** They are much better developed in an adult and may not be seen in the fetus. The passage between the vocal cords is known as the **glottis.**

Trace the trachea caudally. Observe that, at a short distance caudal to the larynx, it is covered by a small, dark, compact, glandular mass, the **thyroid gland.** The thyroid is an endocrine gland producing thyroxin, a hormone needed by a mammal to maintain its high level of metabolism and heat production. The human thyroid gland is located farther forward on the ventral surface of the larynx. Notice that the trachea is held open by rings of cartilage that are incomplete dorsally. This permits the free passage of air. The **esophagus** is the collapsed tube dorsal to the trachea; in life, it is pushed open by the food that is being swallowed.

B. THE BODY CAVITY

1. Opening the Body Cavity

Expose the organs in the thoracic and abdominal cavities by making the incisions through the body wall shown in Figure 3-4. First continue the incision that you made in the neck (incision 1) caudally to the umbilical cord. Cut all the way through the body wall with a pair of scissors, but lift the body wall toward you as you do so, to prevent cutting any internal organs. Continue caudally by making a pair of incisions, each lateral to the umbilical cord and caudal mammary papillae (incision 2). The midventral strip of tissue lying between this pair of incisions contains the umbilical arteries, urinary bladder, and, in the male, the penis. This strip of tissue can be reflected, or turned back, by parting incision 1 and cutting the **umbilical vein** that extends cranially from the umbilical cord to the liver. Cut the vein in such a way that you will be able to find it again. Look into the abdominal cavity and notice the muscular **diaphragm** that forms a more or less transverse partition between the abdominal and thoracic cavities. Make lateral incisions through the body wall (incision 3) just caudal to the attachment of the diaphragm. Follow the attachment of the diaphragm to the body wall all the way to the back muscles. Next cut through the diaphragm peripherally where it attaches to the body wall. Do this on both sides of the body. Finally, carefully cut through the membranes that bind the thoracic organs to the ventral thoracic wall. You can now bend back the flaps of the body wall, wash out coagulated blood and preserving fluid, and expose the internal organs.

If you are studying the muscular system, the third layer of thoracic musculature (transversus thoracis, Exercise 2) can now be seen on the inside of the thoracic wall near the sternum.

2. The Coelom

The body cavity in which the thoracic and abdominal organs are situated is the **coelom.** It is completely lined with a shiny coelomic epithelium, which also covers all of the internal organs. In life, a bit of fluid within the coelom facilitates the expansion, contraction, and other functional movements of the organs.

Cranial to the diaphragm, the coelom is divided into two lateral **pleural cavities,** which contain the lobed **lungs,** and, between the pleural cavities, a **pericardial sac** containing the **heart** (Figs. 3-5 and 3-6). Cut open

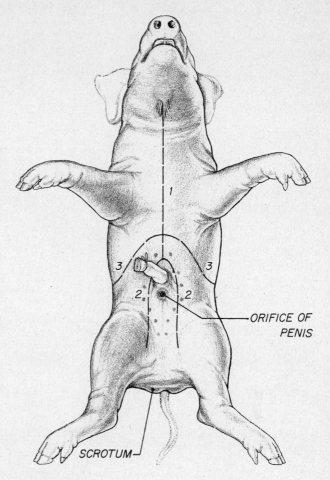

ORIFICE OF PENIS

SCROTUM

FIGURE 3-4
Ventral view of a pig, showing incisions that should be made to open the chest and abdomen.

the pericardial sac if this has not been done. Part of the thymus extends into the thorax and lies on the cranioventral surface of the pericardial sac. The coelomic epithelium in these regions is referred to as pleura and pericardium. **Parietal pleura** lines the chest wall and forms the median wall of each pleural cavity, **visceral pleura** covers the lungs, **parietal pericardium** supported by connective tissue forms the wall of the pericardial sac, and **visceral pericardium** covers the heart. Much of the parietal pleura forming the median walls of the pleural cavities is tightly bound to the parietal pericardium, but it is important to recognize that there are two layers of coelomic epithelium here. The potential space between the two pleural cavities, which is nearly filled by the pericardial sac and thymus, is called the **mediastinum.**

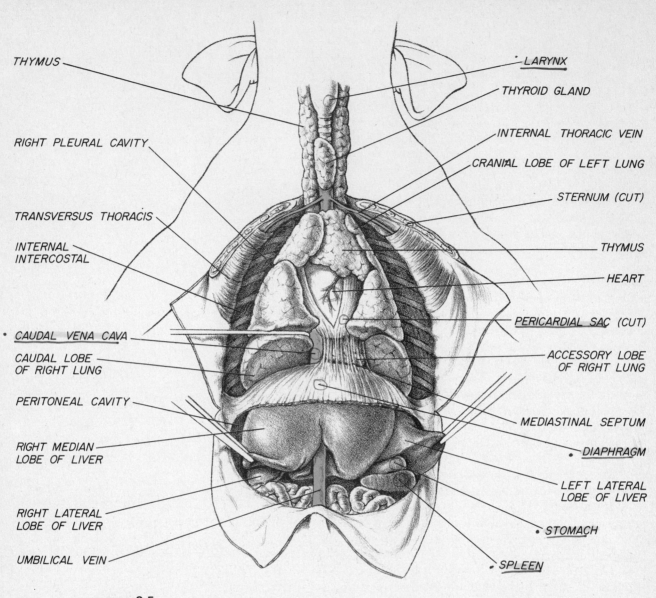

THYMUS

RIGHT PLEURAL CAVITY

TRANSVERSUS THORACIS

INTERNAL INTERCOSTAL

CAUDAL VENA CAVA

CAUDAL LOBE OF RIGHT LUNG

PERITONEAL CAVITY

RIGHT MEDIAN LOBE OF LIVER

RIGHT LATERAL LOBE OF LIVER

UMBILICAL VEIN

LARYNX

THYROID GLAND

INTERNAL THORACIC VEIN

CRANIAL LOBE OF LEFT LUNG

STERNUM (CUT)

THYMUS

HEART

PERICARDIAL SAC (CUT)

ACCESSORY LOBE OF RIGHT LUNG

MEDIASTINAL SEPTUM

DIAPHRAGM

LEFT LATERAL LOBE OF LIVER

STOMACH

SPLEEN

FIGURE 3-5
Ventral view of organs in the cranial thorax and the cranial part of the peritoneal cavity.

The part of the coelom caudal to the diaphragm is the **peritoneal cavity,** and the coelomic epithelium in this region is referred to as the peritoneum. **Parietal peritoneum** lines the body wall, and **visceral peritoneum** covers the abdominal organs (Fig. 3-7). Notice that the organs are, to some extent, united with each other and with the body wall, especially dorsally, by thin membranes called **mesenteries.** Each mesentery consists of two layers of peritoneum between which are connective tissue and blood vessels and nerves passing to the organs.

C. DIGESTIVE AND RESPIRATORY ORGANS OF THE THORAX

The trachea extends caudally and divides into branches, the **bronchi,** that go to the lungs; the bronchi are shown in Figure 3-6, but you will not see them in your specimen until after the heart and its vessels have been removed (Exercise 4). Notice that the left lung is divided into three lobes (cranial, middle, and caudal). The right lung has a fourth accessory lobe that passes dorsal to a large vein (**caudal vena cava**) and lies directly caudal to the heart. Human beings do not have an accessory lobe.

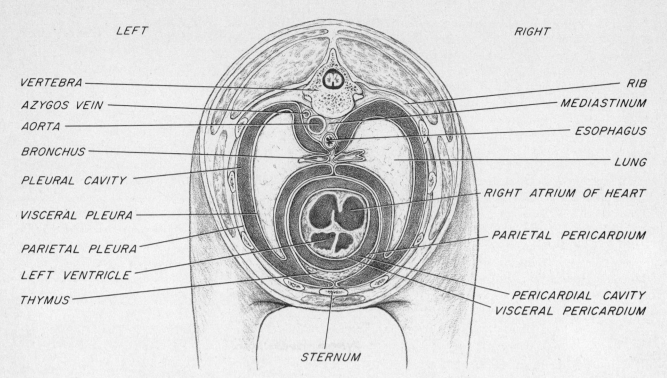

LEFT   RIGHT

VERTEBRA — RIB

AZYGOS VEIN — MEDIASTINUM

AORTA — ESOPHAGUS

BRONCHUS — LUNG

PLEURAL CAVITY — RIGHT ATRIUM OF HEART

VISCERAL PLEURA — PARIETAL PERICARDIUM

PARIETAL PLEURA —

LEFT VENTRICLE — PERICARDIAL CAVITY

THYMUS — VISCERAL PERICARDIUM

STERNUM

FIGURE 3-6
Diagrammatic cross section through the chest of a fetal pig at the level of the heart. Coelomic epithelium is shown by thin, white lines. Viewed from behind.

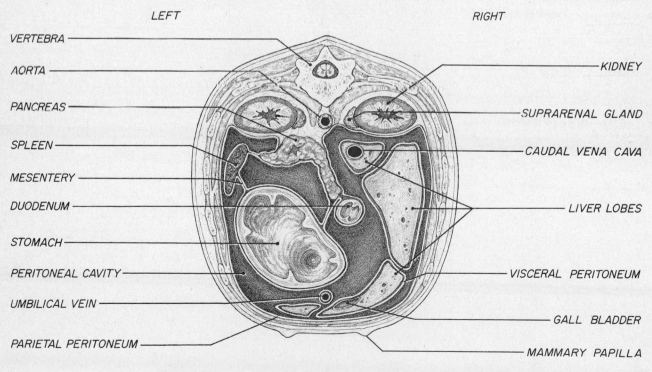

LEFT   RIGHT

VERTEBRA —

AORTA — KIDNEY

PANCREAS — SUPRARENAL GLAND

SPLEEN — CAUDAL VENA CAVA

MESENTERY —

DUODENUM — LIVER LOBES

STOMACH —

PERITONEAL CAVITY — VISCERAL PERITONEUM

UMBILICAL VEIN — GALL BLADDER

PARIETAL PERITONEUM — MAMMARY PAPILLA

FIGURE 3-7
Diagrammatic cross section through the abdomen of a fetal pig at the level of the stomach. Viewed from behind.

Cut off a section of a lung and notice how dense it is. The lungs, of course, have not yet filled with air, nor have they begun to function. After birth, the lung looks a bit more spongy. But all mammal lungs are denser than those of lower terrestrial vertebrates, such as a frog, for the air passages are greatly subdivided, ending in microscopic alveoli that provide a huge surface for the exchange of gases.

Pull the left lung ventrally and toward the right side of the body. Carefully pick away some of the parietal pleura dorsal to it, and you will find the esophagus continuing caudally from the neck. Do not confuse it with the main artery of the body, the **aorta**, which lies against the back. Part of the **vagus nerve** (Exercise 7) can be seen on the surface of the esophagus. Follow the esophagus to the diaphragm, and cut through the diaphragm in order to see where the esophagus enters the stomach.

D. DIGESTIVE ORGANS OF THE ABDOMEN

1. Stomach

The most conspicuous organ in the abdominal cavity is the **liver** (*hepar*) which fits under the dome of the diaphragm. Pull it forward to get a better view of the **stomach** (*ventriculus*), which lies caudal to the liver on the left-hand side of the body (Fig. 3-8). The stomach is a large, saccular, somewhat **J**-shaped organ in which food is stored and in which protein digestion is initiated by the action of the gastric juice containing the enzyme pepsin and hydrochloric acid. Its short border on the right side, extending from the entrance of the esophagus to the beginning of the small intestine, is called the **lesser curvature;** its long, curved border

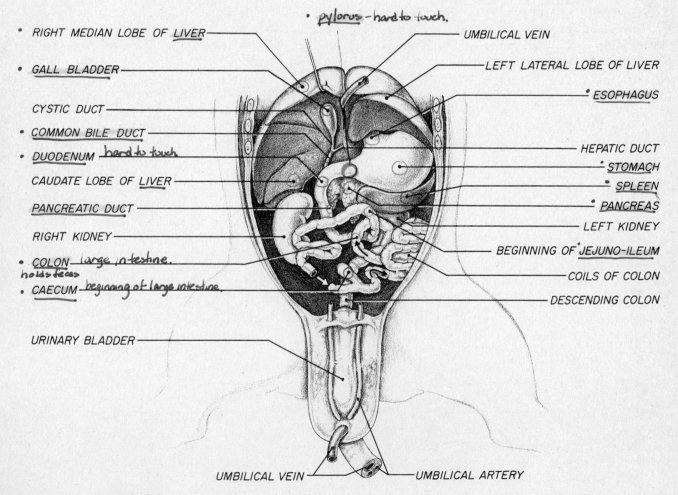

RIGHT MEDIAN LOBE OF LIVER
GALL BLADDER
CYSTIC DUCT
COMMON BILE DUCT
DUODENUM *hard to touch*
CAUDATE LOBE OF LIVER
PANCREATIC DUCT
RIGHT KIDNEY
COLON *large intestine.*
holds feces
CAECUM *beginning of large intestine.*
URINARY BLADDER

pylorus - hard to touch.
UMBILICAL VEIN
LEFT LATERAL LOBE OF LIVER
ESOPHAGUS
HEPATIC DUCT
STOMACH
SPLEEN
PANCREAS
LEFT KIDNEY
BEGINNING OF JEJUNO-ILEUM
COILS OF COLON
DESCENDING COLON

UMBILICAL VEIN
UMBILICAL ARTERY

FIGURE 3-8
Ventral view of abdominal portion of the digestive tract and associated organs of a fetal pig. The liver has been pulled forward and most of the jejuno-ileum has been removed.

on the left side is the **greater curvature.** A mesenteric sac, the **greater omentum,** extends from the greater curvature to the dorsal body wall. The **spleen** (*lien*), an organ that helps produce red blood cells in the fetus and stores and eliminates them in the adult, is an elongate body lying in a part of the greater omentum going to the left side of the stomach. Another mesentery, the **lesser omentum,** extends from the lesser curvature and small intestine to the liver. The liver is also connected by peritoneum to the diaphragm and by the umbilical vein to the ventral body wall. Peritoneum associated with the vein constitutes the **falciform ligament.**

Cut open the stomach, and wash it out. The greenish debris found here and elsewhere in the digestive tract is **meconium.** It consists of bile-stained mucus and sloughed-off epithelial cells of the skin and the lining of the digestive tract. During fetal life some of this material is discharged into the amniotic fluid surrounding the embryo, from which it reenters the digestive tract. It is all discharged in the first, greenish bowel movements of the newborn. Notice that the stomach wall is particularly thick just cranial to the stomach-intestine junction and that its lumen is partly occluded by a fold. This region is called the **pylorus,** and its thick muscular wall forms a sphincter valve that keeps food in the stomach until it is broken down sufficiently to be handled by the intestine. A less conspicuous **cardiac valve** between the stomach and esophagus normally prevents food from going back into the esophagus. You may also notice a small, semi-isolated diverticulum of the stomach near the esophagus. This is believed to be a rudiment of the complex chambering of the stomach seen in cows and other cud-chewing artiodactyles.

2. Small Intestine and Digestive Glands

The first part of the small intestine is the **duodenum.** Trace it, carefully tearing away the attached mesenteries, but not breaking any blood vessels. It extends caudally, crosses to the left side of the body, and ascends toward the stomach. The duodenum is considered to end where it next turns caudally. The rest of the small intestine constitutes the **jejuno-ileum.** There are microscopic intestinal glands that produce enzymes responsible for the final breakdown of food products throughout the small intestine. Cut open a section of the small intestine, wash it out, and notice its velvety lining. This lining consists of numerous microscopic projections, **villi,** that greatly increase the absorptive surface. The membrane supporting most of the small intestine is known as the **mesentery** proper.

The **pancreas** lies in the loop of the duodenum. Its head is in contact with the section of the duodenum leaving the stomach, and a long tail extends across the body toward the spleen. The pancreas secretes enzymes that act upon all major categories of food (carbohydrates, proteins, lipids, and nucleic acids); it also contains endocrine patches (the islets of Langerhans) that produce insulin and glucagon, hormones essential for normal carbohydrate metabolism. The enzymes of the pancreas pass to the duodenum through a minute **pancreatic duct,** which may be found by carefully removing peritoneum from the caudal end of a little tongue of pancreatic tissue that follows along the descending part of the duodenum. Proteolytic enzymes secreted by the stomach and pancreas are not activated until they reach the lumen of the digestive tract.

The liver is divided into five lobes (Figs. 3-5 and 3-8). Notice the **gall bladder** (*vesica fellea*) imbedded in the caudal face of the lobe just to the right of the point at which the umbilical vein enters the liver. Carefully remove the peritoneum from the base of the gall bladder and from the lesser omentum. A **cystic duct** leads out of the gall bladder and joins **hepatic ducts** (one of which is rather prominent) from the liver to form a **common bile duct** (*choledochal duct*) that enters the duodenum just beyond the pylorus. A sphincter at the duodenal end of the common bile duct is normally closed, so that bile secreted by the liver backs up into the gall bladder. When food enters the duodenum, this sphincter is relaxed and bile is discharged into the intestine. Bile contains no digestive enzymes, but it is quite alkaline and neutralizes the acid contents of the stomach, thereby helping to create an environment favorable for the action of pancreatic and intestinal enzymes. Bile also contains salts that emulsify fats and aid in their absorption and pigments derived from the breakdown of hemoglobin. Aside from producing bile, the liver plays a vital role in the storage and metabolism of food products.

3. Large Intestine

Follow the coils of the small intestine, noticing that they are supported by the mesentery, which contains many small blood vessels and lymph nodes. Eventually the small intestine enters the large intestine, or **colon.** It enters on one side in such a way that a short blind sac, the **caecum,** is formed. The caecum is very long in rodents and horses and lodges a colony of microorganisms that enables these animals to digest cellulose. Human beings have a vermiform appendix that lies at the end of a short caecum. Cut open the colon opposite the small intestine and notice the papillalike **ileocaecal opening.** A sphincter at this point prevents material in the colon from backing up into the small intestine.

Most of the colon of the pig forms a tightly coiled mass with a unique pattern. It spirals up on the outside of the mass, reverses and spirals back inside of it, ascends to loop around the duodenum, and finally descends against the back to enter the **pelvic cavity,** which is the region encased by the pelvic girdle. The mesentery supporting it and supplying it with blood vessels is the **mesocolon.** The **rectum,** or terminal part of the large intestine, lies deep within the pelvic cavity and will be seen in a later dissection. It opens on the body surface through the **anus.**

Certain salts and considerable water are absorbed from the digestive residues that enter the colon. In addition, colic bacteria, which reproduce in abundance in this region, synthesize most of the vitamin K needed by mammals for the production in the liver of certain blood-clotting factors. Colic bacteria also help to digest cellulose, especially in herbivores. This accounts for the exceptionally long colon of many herbivores. Finally, undigested residues, disintegrated bile pigments, and many bacteria are discharged as the feces.

4. Structure of the Small Intestine

Study a slide of a cross section through the small intestine of some mammal, and compare it with Figures 3-9 and 3-10. Beginning at the outside and progressing

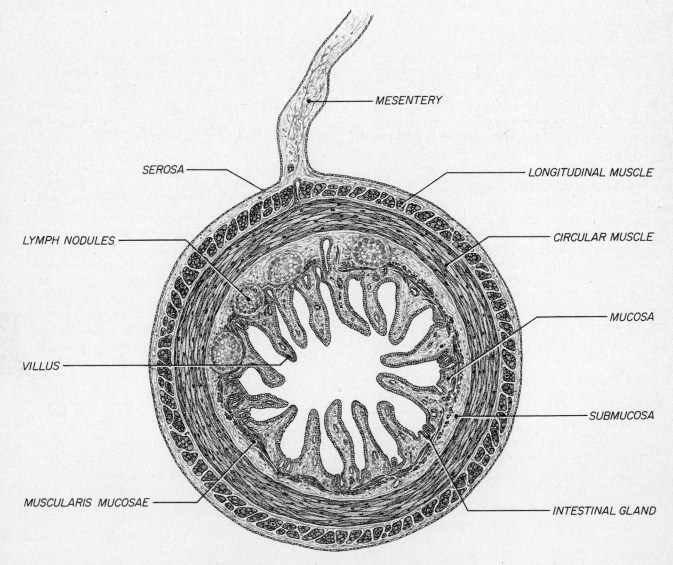

FIGURE 3-9
Microscopic cross section through the ileum of a cat.

toward the lumen, note the following layers: (1) serosa, (2) longitudinal muscle layer, (3) circular muscle layer, (4) submucosa, (5) mucosa.

The **serosa** is a thin layer of simple squamous epithelium (the visceral peritoneum) supported by connective tissue and containing a few blood vessels. About all that can be seen of the epithelial cells are their flattened nuclei. The attachment of the mesentery to the intestine will be seen in some slides.

The **longitudinal muscle layer** is composed of **smooth muscle** fibers that spiral around the intestine in a plane close to its longitudinal axis. Because these are overlapping, spindle-shaped cells being viewed primarily in cross section, their diameters vary considerably, as can be seen with high power. Smooth muscle fibers in the **circular muscle layer** form a tight spiral whose

coils lie close to the transverse plane of the intestine, and so they are seen in lateral view. Notice that, under high power, each of these fibers can be seen to contain a single elongate **nucleus**, located near its center, and many longitudinal, nonstriated **myofibrils**. These two muscle layers, longitudinal and circular, are antagonistic to each other; they produce the churning movements that mix food and enzymes and the peristaltic movements that propel food down the intestine. They are innervated by the autonomic nervous system. The cell bodies of the parasympathetic neurons that activate them appear as clumps of lighter-staining cells between the two muscle layers and in the peripheral part of the submucosa.

The **submucosa** consists of loose fibrous connective tissue containing many blood vessels. If your slide is

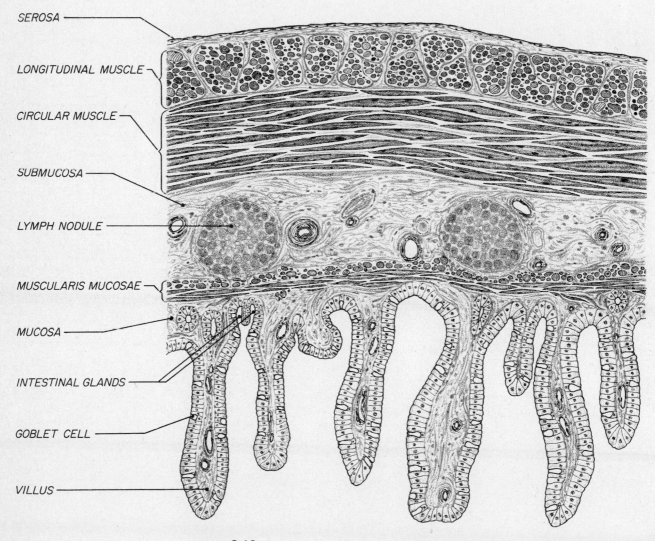

FIGURE 3-10
Enlargement of part of the wall of the ileum of a cat.

taken near the colic end of the small intestine, large, deep-staining masses of lymphocytes called **lymph nodules** will be seen along one side of the intestine, lying in the submucosa and extending into the mucosa. As already seen, T-lymphocytes mature in the thymus. It is probable that a second type, B-lymphocytes, mature in lymphoid tissue in the wall of the digestive tract before distribution to other lymphoid tissues. On exposure to foreign antigens, primarily of bacterial origin, certain B-lymphocytes transform into antibody-producing cells.

The **mucosa** is the most complex layer. A thin layer composed of both longitudinal and circular smooth muscle, the **muscularis mucosae,** lies at its base. Next is a specialized connective tissue containing tubular **intestinal glands** that have invaginated from the simple columnar epithelium that lines the intestine. These glands secrete the intestinal enzymes, some of which are released and act within the lumen of the intestine; others remain associated with microvilli at the surface of the intestinal cells. The numerous fingerlike projections of the mucosa are the **villi,** which extend into the lumen of the intestine. In favorable parts of the slide, intestinal glands can be seen to invaginate between the villi, but in others only the ends of villi and glands are visible. Cells from the muscularis mucosae enter the villi and help to move them. Many roundish, light-staining **goblet cells** will be found in all parts of the epithelium that lines the mucosa. They secrete mucus, which lubricates the food mass and helps protect the delicate intestinal lining.

Circulatory System

The circulatory system is the transport system of the body. In an adult, it carries products absorbed from the digestive tract and oxygen from the lungs to the tissues of the body, and it carries waste products of metabolism to sites of removal. Carbon dioxide is eliminated by the lungs and nitrogenous wastes by the kidneys. In a fetus, all nutrient, respiratory, and excretory exchanges take place through the placenta. The circulatory system also transports hormones, helps to maintain a constant internal environment, and defends the body against disease organisms.

A. ADULT CIRCULATION

The vessels to be exposed will mean more to you if, before dissecting them, you understand the basic pattern of circulation in an adult. Blood is transported in arteries from the heart to minute, thin-walled capillaries in the tissues where exchanges of water, nutrients, gases, and excretory products between the blood and interstitial fluid take place. The tissues are drained primarily by veins, which return to the heart, but some interstitial fluid and a few protein molecules that seep out of the vascular capillaries return to the larger veins in lymphatic vessels. Lymphatic vessels are difficult to see in gross dissections, although some of the lymph nodes that lie along their course will be noticed. Lymph nodes are sites for the production of certain white blood cells, some of which can respond to foreign antigens by transforming into antibody-producing cells.

In an adult mammal (Fig 4-1), blood low in oxygen content (venous blood) is returned from most of the body to the **right atrium** of the heart by two major veins and their tributaries. A **cranial vena cava** drains the head, neck, and arms; a **caudal vena cava,** the caudal parts of the body. It is important to recognize that blood returning from the stomach and intestine goes first to capillarylike spaces in the liver (**hepatic sinusoids**) by an **hepatic portal vein.** Many metabolic conversions of absorbed materials occur in the liver. The liver drains into the caudal vena cava through many **hepatic veins.**

As venous blood enters the right atrium, blood high in oxygen content (arterial blood) returns from the lungs to the **left atrium** of the heart in **pulmonary veins.** (Notice that veins are vessels that lead to the heart, regardless of the oxygen content of their blood. Similarly, arteries always lead away from the heart.) When the right and left atria contract, which they do simultaneously, blood within them is discharged into the respective ventricles. When the thick, muscular walls of the ventricles contract, the increasing pressure closes the atrioventricular valves, and so blood is driven with considerable force into the two large arteries leaving the heart. Closure of valves at the base of each of these vessels prevents a backflow of blood into the ventricles as they refill from the atria. A **pulmonary**

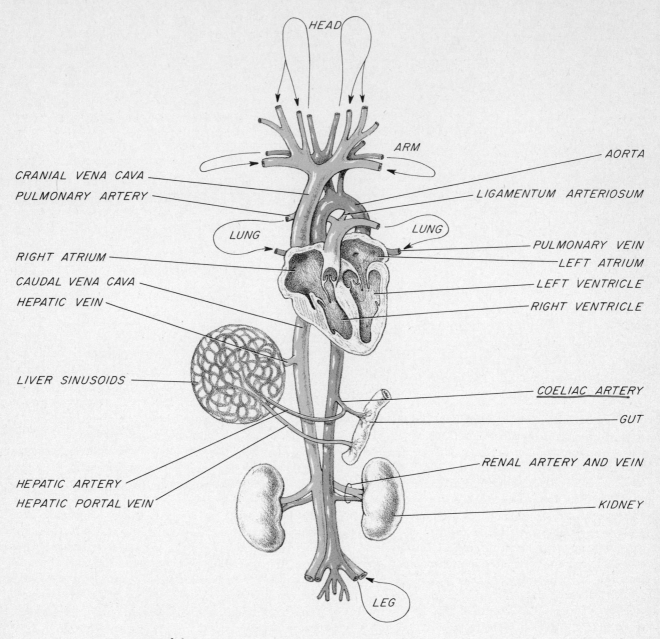

HEAD

ARM

AORTA

CRANIAL VENA CAVA

PULMONARY ARTERY

LIGAMENTUM ARTERIOSUM

LUNG

LUNG

PULMONARY VEIN

RIGHT ATRIUM

LEFT ATRIUM

CAUDAL VENA CAVA

LEFT VENTRICLE

HEPATIC VEIN

RIGHT VENTRICLE

LIVER SINUSOIDS

COELIAC ARTERY

GUT

RENAL ARTERY AND VEIN

HEPATIC ARTERY

KIDNEY

HEPATIC PORTAL VEIN

LEG

FIGURE 4-1

Diagrammatic ventral view of the course of blood flow in an adult mammal.

trunk leaves the **right ventricle** and soon branches into a pair of **pulmonary arteries** that carry the venous blood to the lungs where gas exchange takes place in an adult. An **aorta** leaves the **left ventricle,** arches to the left side of the body, and descends to the pelvis. On its way, it gives off numerous branches that deliver arterial blood to all parts of the body.

Blood flowing through the heart does not supply the heart musculature with needed oxygen and nutrients.

A separate system of **coronary arteries** leaves the base of the aorta and leads to capillary beds in the heart wall. **Coronary veins** drain the heart wall and return venous blood to the right atrium.

Remember, as you dissect your specimen, that it is a fetus. Some aspects of the circulatory pattern will be different than in an adult because the placenta is the site for the exchange of gases, food, and waste products. The difference in pattern between the fetus and adult and the changes that occur at birth will be summarized after you have seen the major blood vessels.

Both arteries and major veins should have been injected in your specimen, but valves in the veins sometimes prevent the thorough penetration of the material injected. If a vein is uninjected, it will appear as a thin-walled, translucent tube in which a bit of blood or injection fluid can be seen. Because most veins lie beside corresponding arteries, it will be convenient to study the arteries and veins together in many regions of the body. Do not be surprised to find variation among individual specimens in the locations at which smaller vessels join the main ones. Often the only way to identify a vessel is to trace it to the organ it supplies or drains.

B. THE HEART AND ITS GREAT VESSELS

Carefully remove the thymus from the base of the neck and the front of the thorax, and cut away the pericardial sac. The heart (Fig. 4-2) consists of a **right** and a **left ventricle**, which can often be distinguished externally on the ventral surface of the heart by a groove containing a prominent coronary artery and vein, and a **right** and a **left atrium**, whose ear-shaped, dark-colored **auricles**, which lie on each side of the front of the heart, are conspicuous in a ventral view.

The large vein that you see entering the right atrium cranially is the **cranial,** or **superior, vena cava.** Its trib-

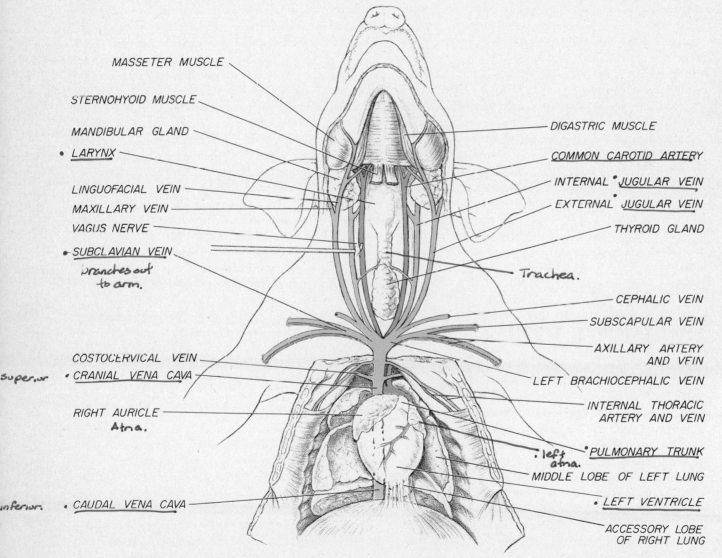

MASSETER MUSCLE
STERNOHYOID MUSCLE
MANDIBULAR GLAND
• LARYNX
LINGUOFACIAL VEIN
MAXILLARY VEIN
VAGUS NERVE
• SUBCLAVIAN VEIN
branches out to arm.
COSTOCERVICAL VEIN
Superior • CRANIAL VENA CAVA
RIGHT AURICLE
Atria.
inferior • CAUDAL VENA CAVA

DIGASTRIC MUSCLE
COMMON CAROTID ARTERY
INTERNAL • JUGULAR VEIN
EXTERNAL • JUGULAR VEIN
THYROID GLAND
Trachea.
CEPHALIC VEIN
SUBSCAPULAR VEIN
AXILLARY ARTERY AND VEIN
LEFT BRACHIOCEPHALIC VEIN
INTERNAL THORACIC ARTERY AND VEIN
• *left atria.* • PULMONARY TRUNK
MIDDLE LOBE OF LEFT LUNG
• LEFT VENTRICLE
ACCESSORY LOBE OF RIGHT LUNG

FIGURE 4-2
Ventral view of the heart and cranial veins and arteries after removal of the thymus and superficial neck muscles. Veins have been colored blue and arteries red, to aid in distinguishing them, but in a fetus many of these vessels carry a mixture of arterial and venous blood.

utaries drain the head, neck, pectoral appendage, and cranial parts of the chest. A **caudal,** or **inferior, vena cava** perforates the diaphragm and enters the right atrium caudally. Its tributaries drain the caudal parts of the body, and in the fetus the placenta as well.

Right atrial blood goes into the right ventricle, which pumps it to the lungs in the **pulmonary trunk.** The pulmonary trunk leaves the cranial end of the right ventricle and curves dorsally between the auricles. By carefully dissecting between it and the left atrium, find where it bifurcates into the two **pulmonary arteries,** which continue to the lungs. In the fetus, part of the pulmonary trunk also continues as a **ductus arteriosus** to join the aorta. Much of the fetal blood bypasses the lungs in this way.

Blood is aerated in the lungs of an adult and returns to the left atrium in **pulmonary veins.** Pulmonary veins can be seen leaving the base of each lung, but their point of entrance into the dorsal surface of the left atrium can be seen more clearly when the heart has been removed (Fig. 4-7).

Left atrial blood enters the left ventricle, which pumps it into the **aorta.** The aorta leaves the heart dorsal to the pulmonary trunk and then emerges on the right side of the pulmonary trunk before it arches to the left side of the body and turns caudally. The first branches of the aorta are a pair of small **coronary arteries,** which arise from the base of the aorta deep to the pulmonary trunk. The more conspicuous left one passes diagonally across the ventral surface of the heart between the right and left ventricles. Both carry blood to capillary beds in the heart wall.

C. VEINS AND ARTERIES OF THE THORAX, NECK, AND HEAD

Cut through muscles that attach to the front of the sternum (sternomastoid and sternohyoid), and finish removing the thymus. Trace the cranial vena cava forward, noticing first its major tributaries (Fig. 4-2). The cranial vena cava is formed near the base of the neck by the confluence of a pair of large **brachiocephalic veins.** The brachiocephalic veins are very short and each, in turn, is formed by the union of a **subclavian vein,** draining most of the shoulder and front leg, and two jugular veins, draining primarily the head and neck. The **external jugular vein** is the more lateral and superficial. It normally receives tributaries from the shoulder in addition to those from the head and neck. The **internal jugular vein** is the more medial. It lies close to the trachea and is accompanied by the **internal carotid artery.** Sometimes the two jugular veins unite with each other before joining the sub-

clavian. Because the subclavian vein extends laterally just cranial to the first rib, it becomes the **axillary vein** when it passes by the axilla, or armpit, and the **brachial vein** when it enters the upper arm. This vessel can be distinguished from others because it lies deep within the upper arm beside the **brachial artery** and several prominent nerves extending into the arm.

Several tributaries of certain of these vessels may be seen, but the pattern will be somewhat variable. A **cephalic vein** lies just beneath the skin on the flexor side of the upper arm and typically enters the base of the external jugular. A **subscapular vein** drains the medial side of the scapula and usually enters the axillary vein.

If your specimen is well injected, certain smaller tributaries of the cranial vena cava also can be found. A pair of **internal thoracic veins** enters the ventral surface of the vena cava shortly caudal to its origin. They extend along the lateral border of the sternum, accompanied by **internal thoracic arteries,** and help drain the chest wall. A pair of **costocervical veins** can be found at about the same level, but deeply because they enter the dorsal surface of the venal cava. They drain parts of the neck, back, and cranial intercostal spaces, and they are accompanied by **costocervical arteries.**

In order to see the arteries that accompany the veins more clearly, cut through the left brachiocephalic and costocervical veins and push the veins aside. As the aorta arches toward the left side of the body (Fig. 4-3), it gives off the **brachiocephalic trunk** and the **left subclavian artery.** Trace the brachiocephalic. It soon branches into a **right subclavian artery** and a pair of **common carotid arteries.** Each subclavian artery gives rise to a **costocervical trunk** and an **internal thoracic artery,** which accompany the corresponding veins. As the subclavian artery curves laterally to enter the axilla and arm, it gives off a **thyrocervical trunk,** which supplies the thyroid gland and parts of the neck. The common carotid arteries extend cranially with the internal jugular vein. At the base of the head, the common carotid divides into **external** and **internal carotid arteries** which respectively supply the superficial and deeper parts of the head.

Dissect between the common carotid artery and the internal jugular vein. The white strand between them is the **vagus nerve,** a part of the autonomic nervous system that supplies most of the viscera with parasympathetic fibers (Exercise 7). A narrower **sympathetic cord,** also a part of the autonomic nervous system, lies deep to the vagus. Trace these nerves caudally. The vagus passes superficial (i.e., ventral) to a subclavian artery and extends to the esophagus where you have previously seen it (Exercise 3). The sympathetic

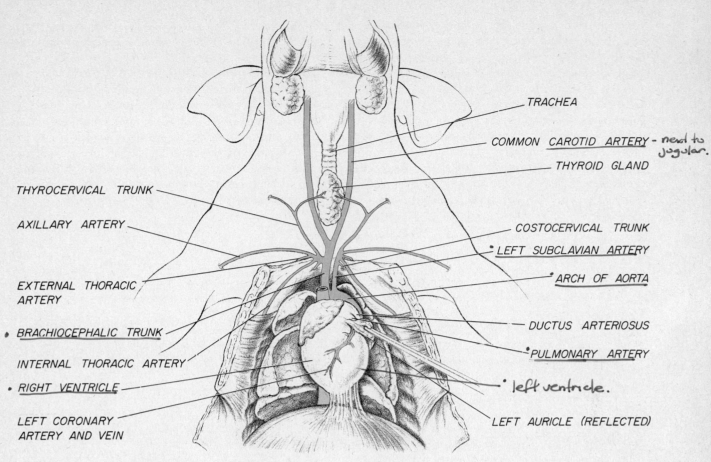

TRACHEA

COMMON CAROTID ARTERY — *next to jugular.*

THYROID GLAND

THYROCERVICAL TRUNK

AXILLARY ARTERY

COSTOCERVICAL TRUNK

*LEFT SUBCLAVIAN ARTERY

*ARCH OF AORTA

EXTERNAL THORACIC ARTERY

BRACHIOCEPHALIC TRUNK

DUCTUS ARTERIOSUS

*PULMONARY ARTERY

INTERNAL THORACIC ARTERY

RIGHT VENTRICLE

*left ventricle.

LEFT CORONARY ARTERY AND VEIN

LEFT AURICLE (REFLECTED)

FIGURE 4-3
Ventral view of the heart and cranial arteries after removal of the veins.

cord goes deep (dorsal) to a subclavian artery and will be seen later on the inside of the chest wall near the point of attachment of the ribs to the vertebral column (Exercise 7).

After giving rise to the left subclavian artery, the aorta curves caudally and descends along the dorsal thoracic wall. Pull the left lung ventrally and follow the aorta. It soon receives the ductus arteriosus from the pulmonary trunk (Fig. 4-3) and gives rise to many pairs of **intercostal arteries** that supply most of the intercostal muscles between the ribs. **Intercostal veins** drain the intercostal spaces. Those from both the left and the right side of the body collect to form a **left azygos vein** that turns ventrally across the aorta and then crosses the dorsal surface of the heart (Fig. 4-6) to enter the right atrium beside the entrance of the caudal vena cava. As the left azygos crosses the heart, it receives several **coronary veins** from the heart wall.

The cranial arteries and veins are very similar in human beings. Major differences are the origin of the left common carotid artery from the arch of the aorta and the presence of both left and right azygos veins.

D. VEINS AND ARTERIES CAUDAL TO THE DIAPHRAGM

1. Vessels on the Dorsal Abdominal and Pelvic Walls

Cut through the left side of the diaphragm, push the abdominal viscera to the specimen's right, and trace the **aorta** into the abdominal cavity. It passes caudally between the two **kidneys** accompanied by the **caudal vena cava** (Fig. 4-4). The narrow, band-shaped organ that lies against the craniomedial border of each kidney is a **suprarenal gland.** The medullary part of this endocrine gland secretes epinephrine, or adrenalin, a hormone that helps adjust the body to stress, and the cortical part secretes many hormones that participate in salt and mineral metabolism, the conversion of protein into carbohydrate, and the development of male secondary sex characters. Women with an overactive suprarenal cortex may, as one symptom, develop a beard and other masculine features.

Carefully clear the surface of the aorta between the

diaphragm and the kidney. You may have to remove the left suprarenal gland and strong, white, fibrous strands that are closely applied to the surface of the vessel. These are **sympathetic nerves** which are derived from the sympathetic cord and follow blood vessels to the visceral organs (Exercise 7).

As the aorta perforates the diaphragm, it gives rise to a median **coeliac artery** that supplies the stomach, spleen, and other cranial abdominal viscera. Shortly caudal to this, another median vessel, the **cranial mesenteric artery,** leaves to go to most of the small intestine. Trace these vessels later. Caudal to them, the aorta gives rise to a pair of large **renal arteries** that supply the kidneys. Paired **renal veins** accompany the arteries and enter the caudal venal cava.

The kidneys lie dorsal to the parietal peritoneum, a position termed retroperitoneal. Lift up the lateral edge of the left kidney and dissect deep to it. You will soon see the **cranial abdominal artery** and **vein** that supply the back in this region and also give rise to small **suprarenal vessels.** The cranial abdominal arteries arise from the aorta. The left cranial abdominal vein normally enters the renal vein; the right one enters the vena cava.

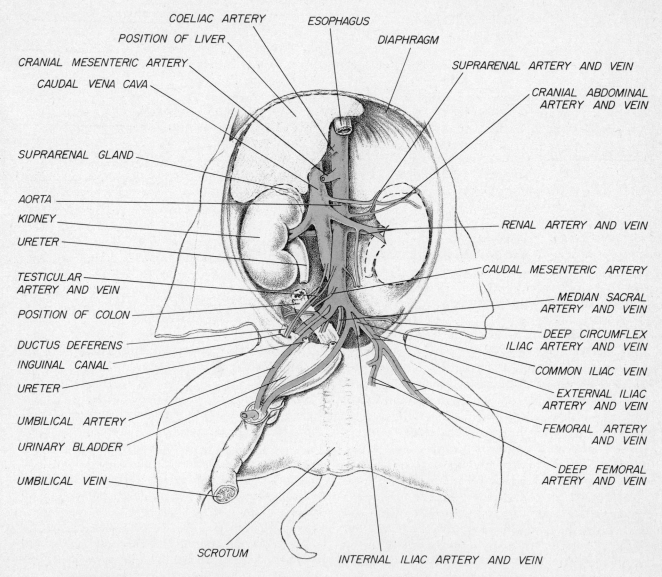

FIGURE 4-4

Ventral view of arteries and veins lying against the dorsal abdominal and pelvic walls. The drawing has been made as if the digestive organs, ureters, and left kidney have been removed, but these organs should only be pushed aside during the dissection.

Notice the **urinary bladder,** which lies against the midventral strip of body wall previously reflected, and find the reproductive organs. If your specimen is a male, the **testes** will have descended into a skin sac, the scrotum, through a passage in the body wall known as the **inguinal canal.** Do not dissect the scrotum and inguinal canal until you study the urogenital system. If you have a female, the paired, coiled **uterine horns** and the **ovaries** can be seen lying deep to the base of the urinary bladder. Small **testicular** or **ovarian vessels** supply the gonads. The paired arteries arise from the aorta caudal to the renal arteries; the veins enter the caudal vena cava, or the left vein may enter the left renal vein.

A median **caudal mesenteric artery** leaves the aorta slightly caudal to the gonadial arteries and supplies much of the colon.

Carefully dissect away connective tissue from the dorsal body wall lateral to the caudal section of the aorta. Also free, but do not destroy, the large convoluted duct (the **ureter**) passing from the kidney to the base of the urinary bladder. Dorsal to the ureter, the aorta gives rise to a pair of **external iliac arteries.** Trace one. A lateral branch, the **deep circumflex iliac artery,** soon leaves to supply some of the pelvic muscles. It is accompanied by the **deep circumflex iliac vein.** The external iliac artery, now accompanied by the **external iliac vein,** continues caudally. After perforating the muscle layers of the body wall, the vessels enter the leg as the **femoral artery** and **vein.** You must separate the muscle layers of the abdominal wall from the skin over the caudal abdominal wall and thigh to see these vessels clearly. A **deep femoral artery** and **vein** leave the median side of the proximal end of the femoral vessels.

Just caudal to the external iliac arteries, the aorta appears to bifurcate, forming two large vessels that extend onto the ventral body wall beside the urinary bladder and continue into the umbilical cord. The distal parts of these vessels are the **umbilical arteries,** which carry blood to the placenta. The short proximal part of each of these vessels, together with a small dorsal branch extending into the pelvic cavity, is the **internal iliac artery.** (In the adult, this dorsal "branch" and the base of the internal iliac artery are the same size, because the umbilical artery atrophies except for a small part supplying the urinary bladder.) The distal part of the internal iliac artery is accompanied by an **internal iliac vein.** Internal iliac and external iliac veins unite to form a short **common iliac vein,** which joins the caudal vena cava. The terminal branch of the aorta is a small **median sacral artery** that extends caudally along the dorsal pelvic wall and enters the tail as the **caudal**

artery. It is accompanied by a **caudal** and **median sacral vein,** which usually enters one of the common iliac veins. These will be seen more clearly after studying the urogenital organs (Exercise 5).

The pattern of these vessels is similar in human beings except that the external and internal iliac arteries arise from a common iliac artery rather than independently from the aorta.

2. Vessels of the Digestive Organs

The abdominal digestive organs receive blood through branches of the coeliac and the cranial and caudal mesenteric arteries, whose origins from the aorta have been seen. These organs are drained by veins that collect to form a hepatic portal vein that leads to sinusoids in the liver. (Portal veins are defined as those that lead from capillaries in one organ to capillaries in another rather than directly to the heart.) The important role of the liver in the metabolism and storage of absorbed food products is correlated with this circulatory arrangement.

Break the greater omentum, push the stomach and spleen cranially and the intestines caudally, and carefully dissect away the tail of the pancreas. Although often not injected, the major parts of the hepatic portal system can be seen accompanied by branches of the coeliac and cranial mesenteric arteries (Fig. 4-5). Veins from the spleen (lien) and most of the stomach collect to form a **lienogastric vein** that unites with a large **mesenteric vein** coming from the intestinal region to form the hepatic **portal vein.** As the hepatic portal vein passes forward dorsal to the pylorus, it receives a small **gastroduodenal vein.** The hepatic portal vein continues to the liver in the lesser omentum. As it enters the liver, it is joined by branches of the **umbilical vein** returning blood from the placenta.

The coeliac artery gives off a small **phrenic artery** to the diaphragm and then divides into a lienic and a hepatic artery. The **lienic artery** accompanies the lienic vein to the spleen and most of the stomach; the **hepatic artery** follows the hepatic portal vein to the liver, giving off a **gastroduodenal artery** on the way. Branches of the cranial and caudal mesenteric arteries accompany the mesenteric veins.

Return to the part of the caudal vena cava between the kidneys (Fig. 4-4). Trace it forward through the liver by dissecting away tissue from the right dorsal surface of the liver. The caudal vena cava does not carry blood to the liver; the liver sinusoid receives blood from the hepatic portal system, hepatic artery, and, in the fetus, from the umbilical vein as well. As

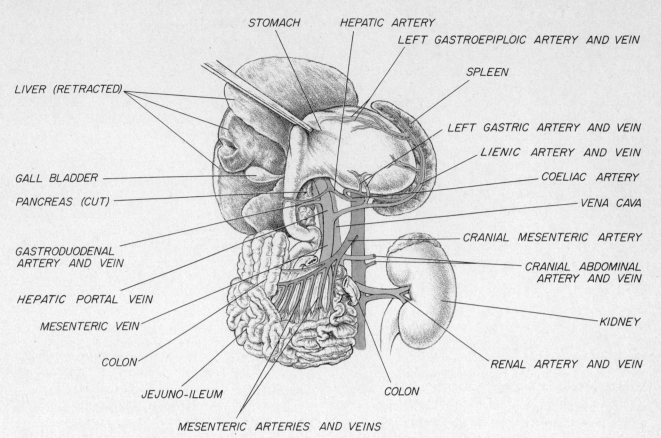

STOMACH HEPATIC ARTERY
LEFT GASTROEPIPLOIC ARTERY AND VEIN
SPLEEN
LIVER (RETRACTED)
LEFT GASTRIC ARTERY AND VEIN
LIENIC ARTERY AND VEIN
COELIAC ARTERY
GALL BLADDER
VENA CAVA
PANCREAS (CUT)
CRANIAL MESENTERIC ARTERY
GASTRODUODENAL ARTERY AND VEIN
CRANIAL ABDOMINAL ARTERY AND VEIN
HEPATIC PORTAL VEIN
KIDNEY
MESENTERIC VEIN
COLON
RENAL ARTERY AND VEIN
JEJUNO-ILEUM COLON
MESENTERIC ARTERIES AND VEINS

FIGURE 4-5
Ventral view of arteries and veins supplying abdominal digestive organs. The stomach and spleen have been pulled cranially, the intestines have been pulled caudally, and the tail of the pancreas has been removed.

the caudal vena cava passes through the liver it receives blood from the liver by many **hepatic veins.** In the fetus some branches of the umbilical vein also join it. You may see one or more **phrenic veins** that drain the diaphragm and enter the vena cava as it goes through the diaphragm into the thorax.

E. LUNGS AND HEART

Carefully cut through the great vessels entering and leaving the heart, and remove it from the body. Before studying the heart, find the trachea at the base of the neck and trace it to the lungs. You can now see the **bronchi** that it gives off to the lungs (Exercise 3). One bronchus arises from the right side of the trachea and passes to the cranial lobe of the right lung. Farther caudally, the trachea bifurcates into two large bronchi, one leading to the rest of the right lung, the other to the left. Trace a bronchus into one lobe of a lung. Notice how it branches and rebranches to form

a respiratory tree whose twigs end in clusters of microscopic air sacs, **alveoli,** in which gas exchange occurs.

The heart creates the hydrostatic pressure that keeps the blood flowing. Its basic structure can be seen quite well in the heart of a fetal pig, but demonstration dissections of an adult sheep or beef heart should also be examined.

The ventral surface of the heart has been observed (Fig. 4-2 and 4-3). Pick away connective tissue to get a clearer view of its dorsal surface (Fig. 4-6). The bifurcation of the pulmonary trunk into the **pulmonary arteries** can now be seen clearly. Just caudal to them, the **pulmonary veins** converge to enter the left atrium. The **right coronary artery** and **vein** course between the right atrium and ventricle and then turn caudally between the left and right ventricles.

Open the heart by making an incision through the lateral wall of each atrium and its auricle (incisions 1 and 2, Fig. 4-7), and carefully remove the injection mass or coagulated blood that it contains. Observe the thinness of the atrial walls. The atria simply collect blood

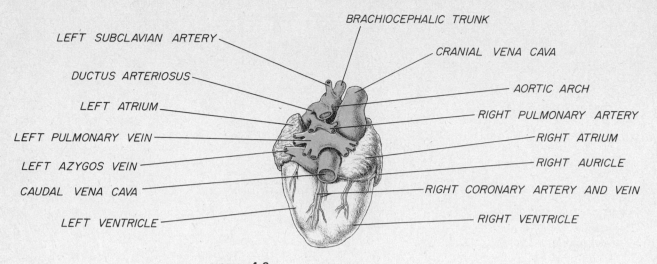

FIGURE 4-6
Dorsal view of the heart and its great vessels.

from the veins during ventricular contraction. During ventricular relaxation, much of the blood is sucked into the ventricles, but their final filling is aided by the contraction of the atria. Notice the entrances of the **cranial** and **caudal vena cava** and **left azygos vein,** which drain the body, into the right atrium and the entrances of the pulmonary veins, which drain the lungs, into the left atrium. At the caudal end of each atrium there is an opening into a ventricle that is guarded by a valve. The valves will be seen more clearly as the dissection progresses.

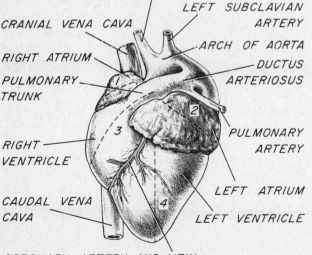
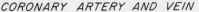

FIGURE 4-7
Ventral view of the heart of a fetal pig showing the location of incisions to be made in opening the heart. (After a drawing by P. Anne Smith.)

Right and left atria are separated from each other by the **interatrial septum,** but in the fetus they are partly connected by the **foramen ovale,** through which some blood bypasses the lungs. A valve in the foramen permits blood to pass only from the right to the left atrium. The entrance to the foramen lies near the dorsal wall of the heart, just cranial to the atrial entrance of the caudal vena cava (Fig. 4-8). Extend the incision that you made through the right atrial wall into the caudal vena cava to see the foramen ovale clearly. The foramen permanently closes at birth but leaves a depression in this region, the **fossa ovalis,** that can be seen in an adult heart.

Again notice the coronary arteries on the dorsal and ventral surfaces of the heart, demarcating the **right** from the **left ventricle.** Cut open the right ventricle by making a diagonal incision through its ventral wall and extending it into the pulmonary trunk (incision 3, Fig. 4-7). Carefully clean out the inside of the ventricle and pulmonary trunk. Notice the thick muscular wall of the ventricle. The three flaps of the tricuspid, or **right atrioventricular,** valve may be seen protruding into the ventricle from the right atrioventricular opening, but they are often torn in removing the injection mass. If so, observe them on a demonstration dissection of a heart. Their margins are attached to the ventricular wall by delicate **tendinous cords.** These permit the flaps to close against each other, yet prevent them from everting into the atrium when the ventricle contracts. Look in the proximal end of the pulmonary artery and notice the three semilunar-shaped pockets of the **pulmonary valve.** This valve prevents blood in the pulmonary trunk from backing up into the ventricle during ventricular relaxation.

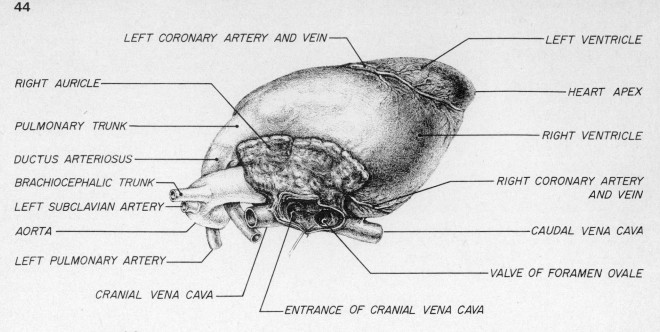

LEFT CORONARY ARTERY AND VEIN

LEFT VENTRICLE

RIGHT AURICLE

HEART APEX

PULMONARY TRUNK

RIGHT VENTRICLE

DUCTUS ARTERIOSUS

BRACHIOCEPHALIC TRUNK

RIGHT CORONARY ARTERY
AND VEIN

LEFT SUBCLAVIAN ARTERY

AORTA

CAUDAL VENA CAVA

LEFT PULMONARY ARTERY

VALVE OF FORAMEN OVALE

CRANIAL VENA CAVA

ENTRANCE OF CRANIAL VENA CAVA

FIGURE 4-8
View of right side of the heart of a fetal pig, with atrium cut open to show the foramen ovale.

Casts of these pockets can often be seen on the injection mass removed from the pulmonary trunk.

Open the left ventricle by making a longitudinal incision through its ventral wall and extending the incision toward the origin of the aorta (incision 4, Fig. 4-7). Clean it out. The muscular wall of the left ventricle is thicker than that of the right. This correlates with the peripheral resistance to be overcome; to all of the body by the left ventricle, only to the lungs by the right ventricle. The bicuspid, or **left atrioventricular, valve** is similar to the right but has only two flaps. The **aortic valve,** which resembles the pulmonary valve, lies in the base of the aorta.

F. FETAL AND NEONATAL CIRCULATION

1. Fetal Circulation

The placenta—rather than the digestive tract, lungs, and kidneys—provides the embryo with food and oxygen and removes carbon dioxide and nitrogenous wastes. Accordingly, the pattern of fetal circulation differs in several important respects from the circulation of an adult (Fig. 4-9). Using the following account of fetal circulation, trace the vessels and identify the parts of the heart on your own specimen.

Blood rich in oxygen and food materials and low in waste products enters the embryo by way of the umbilical vein. Some of this blood joins the hepatic portal system and flows through liver sinusoids, but much

passes directly through the liver in a channel known as the **ductus venosus** into the caudal vena cava and continues on to the right atrium. Some mixing doubtless occurs between the oxygen-rich umbilical blood and the caudal vena caval blood, which is low in oxygen, but this is kept to a minimum by the force of the spurts with which the umbilical blood enters the embryo. Because of the anatomical relationship between the atrial entrance of the caudal vena cava and the foramen ovale, most of the rich umbilical blood bypasses the lungs, passing through the foramen ovale to the left side of the heart. From there it is distributed by the aorta to the head and body. Because branches of the aorta go to the heart wall, the head, and front legs before the aorta receives other blood from the ductus arteriosus, these parts of the fetus receive a particularly rich blood supply.

Most of the blood returning from the head and front legs stays in the right side of the heart because the entrance of the cranial vena cava is directed toward the right atrioventricular opening. This blood, which is low in oxygen content, leaves the heart in the pulmonary trunk going toward the lungs. Throughout much of fetal life, the lungs are not sufficiently develped to handle all of this blood, which is low in oxygen. Moreover, because the lungs are not filled with air, they are compact organs that offer a relatively great resistance to blood flow. Consequently, only a small volume of blood goes through them and returns to the heart in the pulmonary veins. Most of it bypasses the lungs, traveling through the ductus arteriosus to the aorta, where it mixes with rich blood from the left side of the heart. This mixing occurs in a part of the aorta beyond the point at which the important arteries to the head and developing brain have left the aorta. Mixed blood is distributed by the aorta to the rest of the body and by the umbilical arteries to the placenta.

Within the placenta, fetal and maternal blood come so close together that an exchange of food, oxygen, and waste products occurs primarily by diffusion, but there is normally no mixing of the two kinds of blood.

The foramen ovale and ductus arteriosus not only carry a considerable volume of blood around the functionless lungs, but also have another important role. To insure the normal development of their thick, muscular walls, the ventricles must pump a reasonable volume of blood even though little is being pumped to the lungs and little returns from the lungs to be pumped to the body. The ductus arteriosus permits the right ventricle to pump a volume of blood greater than the undeveloped lungs are capable of handling. For this reason the ductus arteriosus is sometimes called the exercise channel of the right ventricle. Similarly, the foramen ovale supplies the left side of the heart with the volume of blood necessary for its development even though little blood returns from the lungs. The foramen ovale is the exercise channel of the left ventricle.

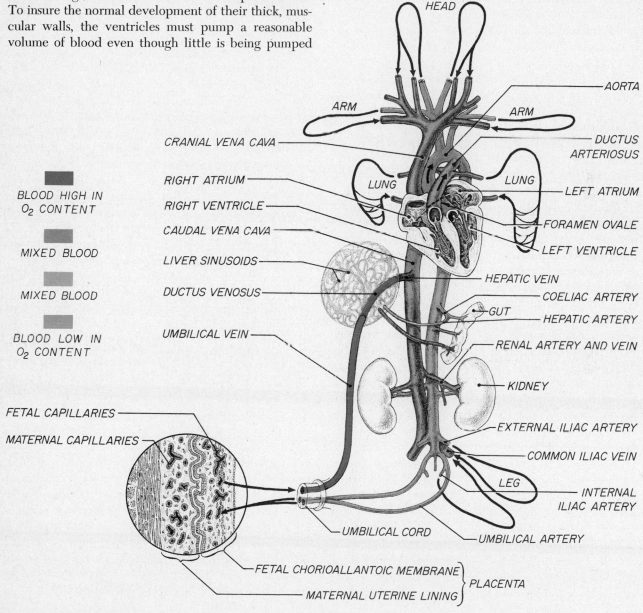

FIGURE 4-9
Diagrammatic ventral view of the course of blood flow in a fetal pig. *Inset:* Microscopic view of part of the placenta.

2. Changes at Birth

In late fetal life, the foramen ovale becomes smaller relative to the rest of the heart and the lumen of the ductus arteriosus becomes relatively narrower, which therefore causes an increasing amount of blood to flow through the lungs. But the adult circulatory pattern is not assumed until after birth. The loss of the placenta at birth decreases the amount of blood returning to the right atrium. Coupled with this there is an increase in the amount of blood returning to the left atrium, because the lungs, now filled with air, offer less resistance to blood flow in the pulmonary circuit. Pressure in the left atrium now equalizes that in the right atrium, so that the flaps of the valve in the foramen ovale remain together. Blood ceases to bypass the lungs by this route.

The ductus arteriosus remains open, but blood flows through it in a direction that is the reverse of the fetal direction. This is because the filling of the lungs with air decreases the resistance in the pulmonary circuit so that it is less than the resistance in the systemic circuit. As a consequence, some of the blood leaving the heart through the aorta (blood that has already been partly aerated by having passed through the lungs) reenters the pulmonary arteries by the ductus arteriosus and passes through the lungs a second time. This double passage of a fraction of the blood through the lungs constitutes the **neonatal circulation.** Blood is more thoroughly aerated at a critical time in life when fetal hemoglobin is being replaced by an an adult type of hemoglobin that has the ability to unload more oxygen in the tissues.

The neonatal pattern lasts only a few hours or a day or so. The ductus arteriosus then contracts and fills in with connective tissue, and the adult circulatory pattern is established. A permanently occluded remnant of the ductus arteriosus remains as the **ligamentum arteriosum** (Fig. 4-1). The flaps of the foramen ovale valve grow together. The site of the foramen ovale is represented in the adult by a depression known as the **fossa ovalis.** The umbilical vein atrophies, as do the parts of the umbilical arteries distal to the urinary bladder.

In a change-over as complex as this, it is not surprising that anomalies sometimes occur. The failure of the ductus arteriosus or foramen ovale to close completely, for example, results in incomplete aeration of the blood—a human infant suffering from this condition is known as a "blue baby."

G. BLOOD

The vessels that you have been dissecting enable blood to be transported between parts of the body. They are important pathways, but it is the blood that performs the transport, defense, and homeostatic functions of the circulatory system. Blood consists of a liquid plasma and several kinds of blood cells carried in the plasma. If human blood smears are available or if you have the material and stains needed to prepare them, the major types of blood cells can be examined.

1. Erythrocytes

Red blood cells, or **erythrocytes,** are the most abundant type of cell, there being about 4.5 to 5.2 million per cubic millimeter of human blood. They appear as pinkish, circular cells about 8 micrometers in diameter (Fig. 4-10A and B). Their nuclei, mitochondria, and some other organelles are lost in the course of their development in mammals, and so they are shaped like biconcave discs. This is most evident in an edge view, but even in a surface view the fact that the center of the cell is thinner than the periphery is usually discernible. This shape gives the cell a particularly large surface area relative to its volume. Erythrocytes are filled with the respiratory pigment **hemoglobin,** which binds reversibly with oxygen, taking it up in the lungs and releasing most of it in the tissues. Most of the carbon dioxide released by the tissues is carried as carbonic acid or its ions and salts, but some also combines with hemoglobin. Because erythrocytes lack nuclei and mitochondria, they live only a few weeks, and they are replaced continually by new ones that develop primarily in the red bone marrow.

2. Leukocytes

The other blood cells are in life colorless or white and are referred to in general as leukocytes. There are several kinds of leukocytes that collectively number from about 5,000 to 9,000 per cubic millimeter of human blood. All are nucleated, and most are larger than the erythrocytes. Some have diameters as large as 20 micrometers. For the purpose of description, it is convenient to divide them into two groups: those having many conspicuous cytoplasmic granules (**glanular leukocytes**) and those having very few cytoplasmic granules or none at all (**agranular leukocytes**).

Lymphocytes are the most common agranular leukocytes comprising from 20 to 25 percent of the leukocyte population. Most are slightly larger than erythrocytes and have a large, nearly spherical nucleus surrounded by only a thin layer of cytoplasm (Fig. 4-10C).

Monocytes are much larger and less common agranular leukocytes that make up from 3 to 8 percent of the white blood cells. The nucleus of a monocyte is typically kidney shaped and the cytoplasm abundant (Fig. 4-10D).

Neutrophiles, which constitute from 65 to 75 percent of all leukocytes, are the most common granular leukocytes. The nucleus is elongated, twisted, and usually constricted into three lobes connected by thin chromatin threads (Fig. 4-10E). Occasionally, part of the nucleus of a female cell is set off as a small appendage called a **"drumstick,"** which consists of the chromatin of the female sex chromosomes. The cytoplasm has a fine, barely perceptible granulation that stains a light purple.

Eosinophiles make up from 2 to 5 percent of the leukocytes. The nucleus usually consists of two lobes connected by a chromatin thread. Many relatively large, reddish granules fill the cytoplasm (Fig. 4-10F).

Basophiles are the rarest of the leukocytes, comprising only 0.5 percent of the population. The elongated and twisted nucleus frequently has an S-like configuration (Fig. 4-10G). The cytoplasm contains bluish granules of different sizes.

Little is known about the functions of leukocytes in the blood stream, but all are capable of amoeboid movement, of squeezing between the endothelial cells of capillary walls, and of wandering in the tissues. They aggregate in areas of infection and inflammation. Lymphocytes, or cells derived from them, have been implicated in the antibody reactions of the body. Neutrophiles are particularly active in engulfing and digesting bacteria. During this process their granules decrease in number, which suggests that the granules are lysosomes containing digestive enzymes. Less is known about the specific functions of other leukocytes, although there is evidence that monocytes transform into tissue macrophages, eosinophiles participate in allergic

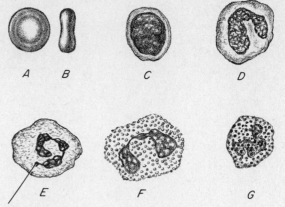

"DRUMSTICK"

FIGURE 4-10
The major types of mammalian blood cells can be distinguished by cell size and shape, nuclear shape, and cytoplasmic granulation: (A) surface view of erythrocyte; (B) edge view of erythrocyte; (C) lymphocyte; (D) monocyte; (E) neutrophile; (F) eosinophile (red granules); (G) basophile (blue granules). (After a drawing by P. Anne Smith.)

reactions, and basophiles release histamine. The relative numbers of the different types present are frequently diagnostic of certain diseases.

3. Blood Platelets

Look for small groups of granules surrounded by bits of cytoplasm that are scattered among the other cells. These are the **blood platelets,** which are simply small blobs of cytoplasm that have budded off certain giant cells (**megakaryocytes**) in the bone marrow. They help protect the body from excess blood loss at the site of an injury: first, by clumping and helping to plug an injured vessel and, second, by releasing **thromboplastin.** Thromboplastin initiates a series of enzyme-mediated reactions that result in the transformation of the soluble plasma protein fibrinogen into fibrin threads that form a blood clot.

Urogenital System

The excretory and reproductive organs of vertebrates are so intimately interrelated—especially in males, in which certain ducts carry both sperm and urine—that it is convenient to regard them together as the urogenital system.

A. EXCRETORY SYSTEM

Carbon dioxide, one waste product of metabolism, is removed from the body by the lungs, but most of the other waste products, and in particular the nitrogenous waste products of protein metabolism, are eliminated by the kidneys. But the kidneys are not just excretory organs. By eliminating a variety of materials that may be present in the body fluids in amounts exceeding the body's needs and conserving those not in excess, the kidneys play an essential role in maintaining an internal environment that is nearly constant in water and salt content, in pH, and in the blood level of sugar and many other substances.

The pair of **kidneys** (*renes*) has been observed, and probably partly uncovered, in previous dissections (Fig. 5-1). Although they bulge into the peritoneal cavity somewhat, they actually lie dorsal to it, against the ventral surface of the back muscles. Coelomic epithelium covers only their ventral surface, and should be peeled off. A narrow, bandlike adrenal, or **supra-**renal, **gland** lies adjacent to the craniomedial border of each kidney (Exercise 4).

Each kidney is drained by a **ureter,** which leaves from the slightly indented medial border of the kidney, accompanied by the renal artery and vein. Trace the ureter caudally, along the muscles of the back; notice that it turns ventrally near the brim of the pelvis to enter the **urinary bladder.** The urinary bladder is attached by a ventral mesentery (the **ventral ligament of the bladder**) to the reflected midventral strip of abdominal wall. Trace the bladder into the umbilical cord and notice that it continues as the **allantoic stalk.** The urinary bladder develops embryonically from the intraabdominal part of the allantois. Most of the allantois extends peripheral to the body as a large extra-embryonic sac that helps to form the placenta (Fig. 5-6). The caudal end of the bladder narrows to form a duct, the **urethra,** that can be seen disappearing into the pelvic cavity. Its subsequent course to the body surface will be seen in the dissection of the reproductive organs.

Return to a kidney, and section it in the frontal plane of the body: that is, cut off its ventral half along the plane of the ureter (Fig. 5-1). Dissect away blood vessels in the half of the kidney left in the body and the ventral wall of the ureter. You will find that, within the kidney, the ureter expands to form a large chamber, the **renal pelvis.** The renal pelvis is partitioned into many smaller compartments, **renal calyces,** each of

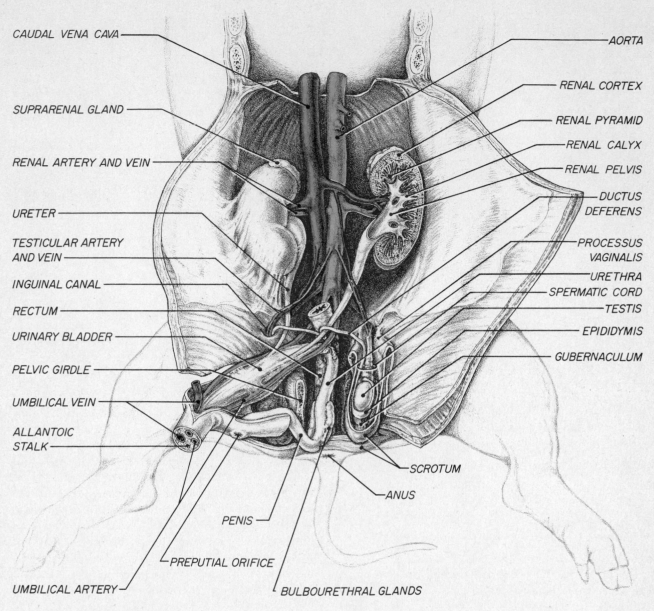

CAUDAL VENA CAVA

AORTA

SUPRARENAL GLAND

RENAL CORTEX

RENAL PYRAMID

RENAL CALYX

RENAL ARTERY AND VEIN

RENAL PELVIS

URETER

DUCTUS DEFERENS

TESTICULAR ARTERY AND VEIN

PROCESSUS VAGINALIS

INGUINAL CANAL

URETHRA

SPERMATIC CORD

RECTUM

TESTIS

URINARY BLADDER

EPIDIDYMIS

PELVIC GIRDLE

GUBERNACULUM

UMBILICAL VEIN

ALLANTOIC STALK

SCROTUM

ANUS

PENIS

PREPUTIAL ORIFICE

UMBILICAL ARTERY

BULBOURETHRAL GLANDS

FIGURE 5-1
Ventral view of the urogenital system of a male pig. The left kidney has been sectioned in the frontal plane.

which receives a dark tuft of kidney tissue. Each of these tufts is a **renal pyramid,** and collectively they constitute the **medulla** of the kidney. The peripheral, lighter-colored part of the kidney is its **cortex.** The kidney is composed of microscopic kidney tubules, or **nephrons,** and associated blood vessels (Fig. 5-2). A simple filtrate of the blood leaves a knot of capillaries known as a **glomerulus** and enters **Bowman's capsules** at the beginning of each nephron. One glomerulus and Bowman's capsule constitute a **renal corpuscle.** These bodies are located in the renal cortex and can sometimes be seen with low magnification in well-injected specimens of sheep or pig kidneys. The glomerular filtrate contains all of the soluble components of the blood in the same proportion as in the blood, except for larger molecules such as plasma proteins. The filtrate then passes in sequence through the **proximal convoluted tubule** in the cortex, the **loop of Henle** that dips into the medulla, the **distal convoluted tubule** in the cortex, and finally a **collecting tubule** that extends through the

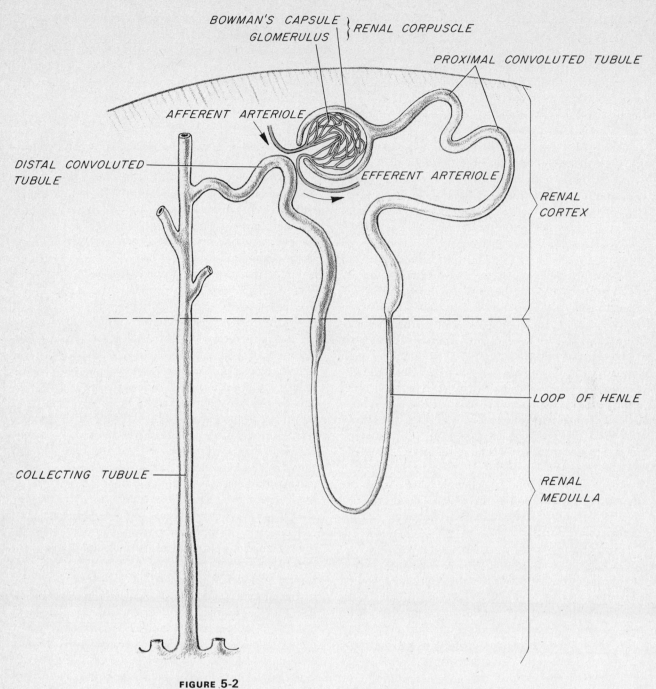

BOWMAN'S CAPSULE
GLOMERULUS } RENAL CORPUSCLE

PROXIMAL CONVOLUTED TUBULE

AFFERENT ARTERIOLE

DISTAL CONVOLUTED TUBULE

EFFERENT ARTERIOLE

RENAL CORTEX

LOOP OF HENLE

COLLECTING TUBULE

RENAL MEDULLA

FIGURE 5-2
Diagram of a nephron and its relation to kidney regions.

medulla to terminate on the surface of the renal pyramids. As the filtrate passes through the nephron there is selective reabsorption of ions, glucose, amino acids, water, and other materials needed by the body into capillaries surrounding the nephron, so that urea and other waste products become progressively concentrated. Sometimes there is selective secretion of additional wastes into parts of the tubule.

B. REPRODUCTIVE SYSTEM

Although you will dissect the reproductive system of only one sex, you should use another student's specimen to study the opposite sex. This means that you should be particularly careful in dissecting your own specimen —as careful as though you were preparing a demonstration preparation, which, in effect, you are.

1. Male Reproductive Organs

At one period of embryonic development, the testes are located dorsal to the coelom and just caudal to the kidneys, but during subsequent development they undergo a caudal migration, or descent, and come to lie in an external pouch, the **scrotum,** part of which can be seen ventral to the anus. Blood vessels, nerves, and the sperm ducts are carried back with them.

Notice again the testicular arteries and veins (Exercise 4) extending caudally to pass through a pair of openings in the abdominal wall (Fig. 5-1). Each opening, an **inguinal canal,** is formed during the descent of a testis. It is not a true perforation of the abdominal wall; rather, it is an evagination of muscular and connective tissue layers of the wall, forming a sac of tissue that becomes a part of the scrotum. Carefully separate the skin from the muscular layers of the abdominal wall caudal to the inguinal canals, and expose the two thin-walled, elongated sacs that extend across the ventral surface of the thigh muscles toward the cutaneous part of the scrotum. Without breaking them open, free the sac on each side of the body from the surrounding structures. In an adult, the wall of the scrotum consists of a cutaneous pouch and the attenuated muscle and connective-tissue layers of the abdominal wall that form the sacs you have exposed. Pass a probe through an inguinal canal, and notice that a coelomic sac, the **processus vaginalis,** extends into the scrotum. In many adult mammals, including male human beings, the proximal part of the processus vaginalis atrophies, but an isolated sac of coelom remains distally around the testis.

Leave the inguinal canal and scrotal sac intact on one side of the body so that you can demonstrate it to a colleague dissecting a female. Open the scrotal sac on the other side, and find the **testis.** A band of tissue, the **epididymis,** starts at the cranial end of the testis, and extends caudally along one side of the testis to its caudal end, where the epididymis runs into the sperm duct, or **ductus deferens.** Sperm is produced in microscopic seminiferous tubules, leaves the testis through microscopic ducts, and enters the cranial end of the epididymis. Most of the sperm is stored in the epididymis until ejaculation. The epididymis and ductus deferens represent part of the ancestral vertebrate kidney system. In fishes and amphibians, sperm pass from the testis to certain kidney tubules, and thence to an archinephric duct that drains the kidneys. This duct thus carries both sperm and urine, though not at the same time. Higher vertebrates have evolved a new kidney duct, the ureter; the old kidney duct, along with a part of the ancestral kidney, has been taken over

completely by the male genital system as the ductus deferens and epididymis.

The cord of tissue that extends from the caudal end of the epididymis to the wall of the scrotum is known as the **gubernaculum.** It extends between the testis and scrotum very early in development, and its failure to grow as rapidly as other parts of the body helps to "pull" the testis caudally. The descent of the testis is related to the high and constant body temperature of mammals. Experiments have shown that in most mammals scrotal temperature is several degrees lower than intraabdominal temperature. The final stage of spermatogenesis, in which the spermatids develop tails and lose most of their cytoplasm (Section C of this exercise), cannot be completed if the testes are confined to the abdominal cavity.

Notice that the testis, epididymis, ductus deferens, and testicular vessels are supported by a mesentery, the **mesorchium,** that extends from the dorsal wall of the processus vaginalis. Slightly cranial to the testis, the testicular artery is highly coiled and entwined by a network of testicular veins known as the **pampiniform plexus.** This is another cooling mechanism in which some heat flows from the warmer arterial blood to cooler venous blood returning from the scrotum before the arterial blood reaches the testis. The ductus deferens, testicular vessels, and an inconspicuous testicular nerve run together between the testis and inguinal canal as a bundle known as the **spermatic cord.** After passing through the inguinal canal, the ductus deferens loops over the ureter and enters the urethra along with the ductus deferens of the opposite side.

The pelvic cavity must be opened to see the rest of the system. But before this is done, locate the **penis** in the midventral strip of body wall that contains the urinary bladder. The **preputial orifice,** just caudal to the umbilical cord, was identified in Exercise 1. The penis extends caudally from here. It may have been seen between the two sacs of the scrotum when they were dissected. Free the penis sufficiently to push it to one side, and cut through the midventral part of the pelvic muscles and the pelvic girdle. Spread the legs apart and expose the pelvic cavity.

Trace the urethra caudally, separating it from the **rectum,** or terminal part of the digestive tract. A pair of small glands, the **seminal vesicles,** lies on the dorsal surface of the urethra at the point at which the deferent ducts enter. Dissect between the seminal vesicles, and find the **prostate** (Fig. 5-3). A larger pair of glands, the **bulbourethral glands,** flank the urethra near the anus. At the time of ejaculation, all of these glands secrete a liquid that carries the sperm: this is the seminal fluid.

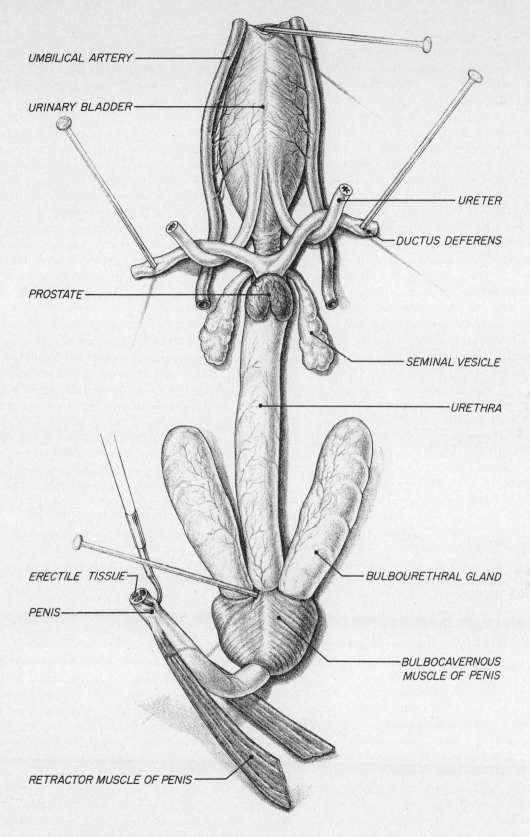

UMBILICAL ARTERY

URINARY BLADDER

URETER

DUCTUS DEFERENS

PROSTATE

SEMINAL VESICLE

URETHRA

BULBOURETHRAL GLAND

ERECTILE TISSUE

PENIS

BULBOCAVERNOUS
MUSCLE OF PENIS

RETRACTOR MUSCLE OF PENIS

FIGURE 5-3
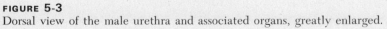
Dorsal view of the male urethra and associated organs, greatly enlarged.

Just distal to the bulbourethral glands, the urethra enters the penis, which extends cranially beneath the skin. The base of the penis is enlarged and surrounded by a **bulbocavernous muscle** that aids in expelling urine and seminal fluid. A slender pair of **retractor muscles of the penis** extends cranially from the base of the penis, each muscle passing on either side of the rectum. Cut open the preputial orifice, and find the distal end of the penis (**glans penis**). The glans penis is very long in mammals such as the pig in which the penis is embedded in the ventral abdominal wall. It lies in a deep recess of the skin known as the penis sheath, or **prepuce.** The retractor muscle of the penis pulls the penis back into its sheath after an erection. Cut a cross section out of the middle of the penis and observe it with low magnification. The penile part of the urethra will be seen to be surrounded by spongy **erectile tissue,** which becomes engorged with blood during an erection. Two other columns of erectile tissue extend along the dorsal surface of the penis. The penis of a man is very similar although its distal part is independent of the ventral abdominal wall, and there is no retractor muscle.

2. Female Reproductive Organs

The **ovaries** are a pair of small nodulelike organs located just caudal to the kidneys. They are supported in a mesentery, the **broad ligament,** in the free edge of which is a conspicuous convoluted duct, the **horn of the uterus** (Fig. 5-4). Another mesentery, the **round ligament,** crosses the broad ligament at right angles and attaches to the abdominal wall near the groin at a point comparable to the location of the male inguinal canal. The round ligament is the female counterpart of the male gubernaculum and helps bring about a partial descent of the ovary.

At the caudal end of the ovary, the uterine horn narrows abruptly to form the small, highly convoluted fallopian tube, or **uterine tube,** that proceeds to the cranial end of ovary. Here it expands to form a little hood known as the **infundibulum,** whose opening into the abdominal cavity (the **ostium tubae**) receives the eggs as they break out of the ovary. Eggs are carried into the ostium by ciliary currents, and the union of sperm and eggs occurs in the uterine tubes.

Trace the uterine horns caudally, and observe that they unite to form the **body of the uterus,** which lies dorsal to the urethra. The pelvic cavity must be cut open to expose the rest of the system. Reflect the skin from the ventral surface of the pelvis and, with a scalpel, cut vertically down along the midventral line

through the pelvic muscles and girdle. Spread the legs apart and open the pelvic cavity. Separate the urethra from the body of the uterus and from the **vagina,** into which the uterus leads. Vagina and urethra unite to form a common passage, the urogenital canal, or **vaginal vestibule,** which leads to the body surface. The opening on the surface, together with the surrounding skin folds (**labia**), constitutes the **vulva.**

Separate the uterus, vagina, and vestibule from the **rectum,** or terminal part of the digestive tract. Cut through the skin between the urogenital orifice and anus, and remove the urogenital organs. Cut open the vestibule, vagina, and uterus on their dorsal surfaces (Fig. 5-5). A small protuberance, the **glans clitoridis,** will be seen on the ventral surface of the vestibule near the base of the external **genital papilla.** It is a little mass of erectile tissue comparable to the male glans penis. Also note the opening of the urethra.

The parts of the uterus and vagina can be seen particularly well in this dorsal dissection of the urogenital tract. The horns of the uterus unite to form the body of the uterus, and this leads into the neck, or **cervix,** of the uterus. The cervix can be distinguished internally by the interdigitating folds that partially occlude its lumen. The smoother part of the passage that lies between the cervix and vestibule is the vagina.

A major difference in the genital system of a woman is the absence of uterine horns. The uterine tubes lead directly to a large, pear-shaped uterine body. Vagina and urethra do not unite posteriorly to form a common tube but open independently into a shallow vaginal vestibule between the genital labia.

3. The Fetal Pig and Its Extraembryonic Membranes

Fertilized eggs are carried by ciliary currents and peristaltic contractions through the uterine tubes into the horns of the uterus where embryonic development is completed in 112 to 115 days. If the uterus of a pregnant sow is available, it can be seen that the fetuses tend to be equally spaced in the two horns and that each brings about a local enlargement of the horn. Litter size normally ranges from 6 to 12.

Open one of the compartments of the horn, and notice that each fetus is enclosed within an elongated, sausage-shaped **chorionic vesicle** (Fig. 5-6). The surface of the vesicle bears many folds, including many that can be seen only with magnification. All of them interdigitate with corresponding folds of the uterine lining. There are also many round bumps, **areolae,** over the surface of the chorion; usually each one is

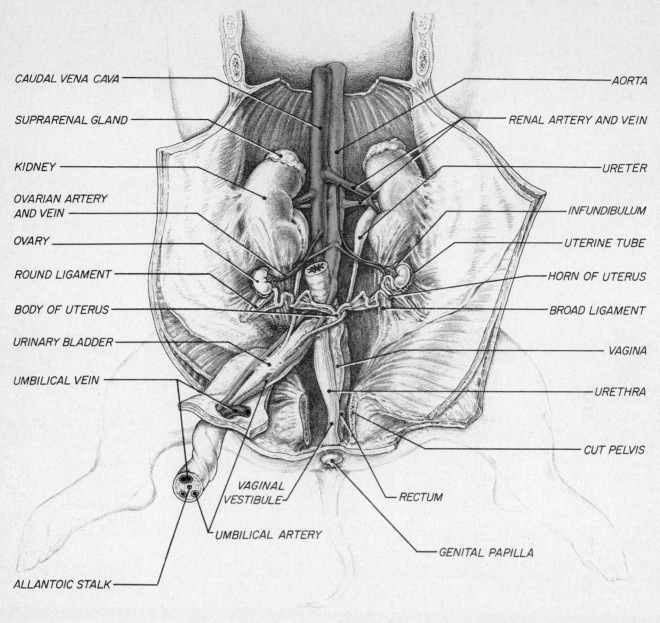

CAUDAL VENA CAVA

SUPRARENAL GLAND

KIDNEY

OVARIAN ARTERY
AND VEIN

OVARY

ROUND LIGAMENT

BODY OF UTERUS

URINARY BLADDER

UMBILICAL VEIN

VAGINAL
VESTIBULE

UMBILICAL ARTERY

ALLANTOIC STALK

AORTA

RENAL ARTERY AND VEIN

URETER

INFUNDIBULUM

UTERINE TUBE

HORN OF UTERUS

BROAD LIGAMENT

VAGINA

URETHRA

CUT PELVIS

RECTUM

GENITAL PAPILLA

FIGURE 5-4
Ventral view of the urogenital system of a female pig.

adjacent to the orifice of a uterine gland. The **placenta** is a combination of uterine lining and the wall of the chorionic vesicle (Fig. 4-9). In a fetal pig, the union is very loose. Food, gases, and waste products diffuse between the fetal and maternal parts of the placenta, crossing the slight space between them, which is filled with a uterine secretion. In many other mammals, including human beings, there is a more intimate union, because parts of the chorionic vesicle penetrate the uterine lining.

Cut open the chorionic vesicle, being careful not to

cut through or break a second sac that lies within it, surrounding the fetus. Most of the wall of the vesicle is composed of a fusion of two extraembryonic membranes that develop in association with the fetus. The outer one is the **chorion,** the inner one the wall of the **allantois.** The allantois is a large sac that grows out from the fetus. Its stalk was seen in the umbilical cord, and the intraembryonic part of it becomes the urinary bladder. When the chorionic vesicle is opened, the cavity of the allantois is exposed. **Umbilical blood vessels,** which can be seen ramifying in the wall of the

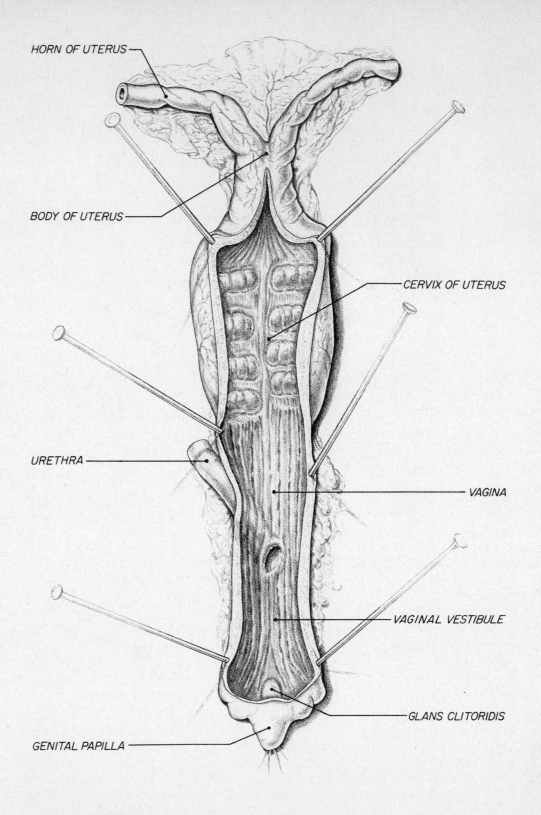

HORN OF UTERUS

BODY OF UTERUS

CERVIX OF UTERUS

URETHRA

VAGINA

VAGINAL VESTIBULE

GLANS CLITORIDIS

GENITAL PAPILLA

FIGURE 5-5
Dorsal view of uterus, vagina, and vaginal vestibule, cut open to show internal appearance, greatly enlarged.

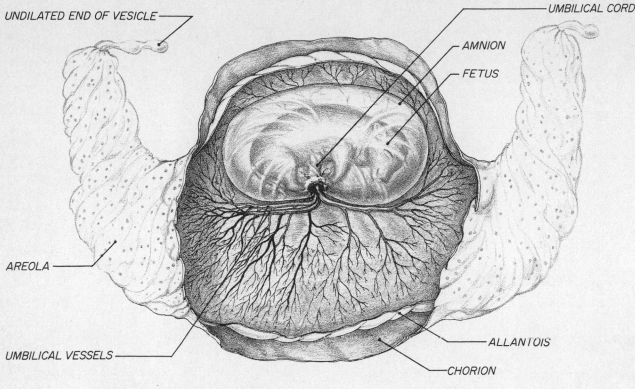

UNDILATED END OF VESICLE

UMBILICAL CORD

AMNION

FETUS

AREOLA

UMBILICAL VESSELS

ALLANTOIS

CHORION

FIGURE 5-6
Opened chorionic vesicle of a pig. The wall of the allantoic sac is fused with the chorion. Part of it is peeled away in this dissection.

chorionic vesicle and entering the **umbilical cord,** lie in the wall of the allantois. The chorion itself is not vascular. You may notice an undilated part of the chorionic vesicle at each end. This is composed only of chorion, for the allantois does not penetrate into the ends of the vesicle.

The fetus is surrounded by a thin-walled, nonvascular **amnion. Amniotic fluid** fills its cavity, prevents adhesions between the fetus and surrounding membranes, acts as a protective water cushion, and is in a sense a local aquatic environment in which the embryo develops. Ancestral vertebrates, of course, developed in water. It was the evolution of the amnion and other extraembryonic membranes that permitted the higher vertebrates (reptiles, birds, and mammals) to reproduce in the terrestrial environment.

C. THE GONADS

The reproductive passages that you have been studying transport gametes and protect and nourish the developing embryo. It is the gonads—the testis and ovary—that produce the gametes. Use slides of mammalian gonads to study gamete production.

1. Testis

The **testis** is composed primarily of many long and highly coiled **seminiferous tubules,** of which some will be seen cut in cross section and others tangentially, or in oblique section (Fig. 5-7). The total length of all of them in a human testis is estimated to be 250 meters! A tough, fibrous capsule, the **tunica albuginea,** forms the wall of the testis and sends septa into the organ in certain regions. A thin **coelomic epithelium** lies peripheral to the tunica.

Examine the walls of several seminiferous tubules. Chromosomes will be visible in many of the cells for they are dividing rapidly. The periphery of the tubule is occupied by **spermatogonia,** which divide mitotically. Some of the daughter cells remain as spermatogonia; others enlarge, move toward the lumen, and become **primary spermatocytes.** Spermatogonia and primary spermatocytes contain the diploid number of chromosomes. Each primary spermatocyte undergoes a meiotic division to form two **secondary spermatocytes,** and each of these divides meiotically again to form two **spermatids.** Thus, four haploid spermatids develop from a single diploid primary spermatocyte. There is no growth of the cells between these divisions,

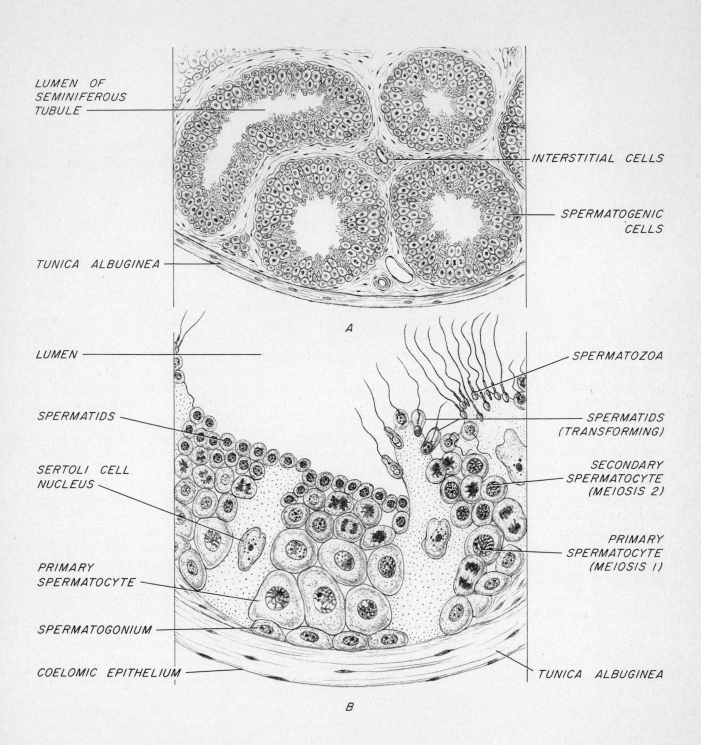

LUMEN OF
SEMINIFEROUS
TUBULE

INTERSTITIAL CELLS

SPERMATOGENIC
CELLS

TUNICA ALBUGINEA

A

LUMEN

SPERMATOZOA

SPERMATIDS

SPERMATIDS
(TRANSFORMING)

SERTOLI CELL
NUCLEUS

SECONDARY
SPERMATOCYTE
(MEIOSIS 2)

PRIMARY
SPERMATOCYTE

PRIMARY
SPERMATOCYTE
(MEIOSIS 1)

SPERMATOGONIUM

COELOMIC EPITHELIUM

TUNICA ALBUGINEA

B

FIGURE 5-7
Diagrammatic microscopic section of part of a rat testis:
(A) low-power view of several seminiferous tubules;
(B) high-power view of part of one seminiferous tubule.

and so they become progressively smaller and move closer to the lumen of the tubule. The stages of spermatogenesis are most easily identified by relative cell size and position within the wall of the tubule.

A spermatid undergoes a transformation into a mature, motile **spermatozoon.** Its nucleus condenses and appears very small and dark, much of the cytoplasm is lost, and a tail develops. During this process it becomes more evident that the spermatogenic cells are closely associated with large, **Sertoli cells,** which extend from the base to the lumen of the tubule between groups of spermatogenic cells. The outlines of Sertoli cells are difficult to see with ordinary stains, but their positions can be discerned by observing the groups of maturing sperm associated with them.

Relatively large, light-staining **interstitial cells** are present in groups in the connective tissue between the seminiferous tubules. They produce **testosterone,** the male hormone responsible for the development of male secondary sex characters. Testosterone production, in turn, is controlled by pituitary **gonadotropic hormones.** Both testosterone and gonadotropic hormones are needed for sperm production. Spermatogenesis is seasonal in seasonally breeding mammals but is continuous during the adult life of human males. Separate groups of spermatogenic cells mature at different times; so the particular combination of types of spermatogenic cells may be different in different tubules or different parts of the same tubule. Some groups may be in early stages of spermatogenesis; others, in the spermatid and spermatozoa stages. It probably takes 64 days for human spermatozoa to develop from spermatogonia, but the area of sperm production is so large that vast numbers are produced, as many as 200 to 300 million per ejaculation.

2. Ovary

A small, highly vascular connective tissue forms the **medulla** of the **ovary,** but most of the organ is a thick, connective-tissue **cortex** containing maturing eggs, each of which is surrounded by follicular cells (Fig. 5-8). The surface of the ovary is covered by a simple, squamous epithelium known as the **germinal epithelium.** The periphery of the cortex is filled with several layers of **primordial follicles.** Each contains a large **primary oocyte** characterized by a conspicuous eccentric nucleus imbedded in relatively clear cytoplasm. A single layer of flattened follicle cells, known as **granulosa cells,** surrounds the primary oocyte. The primary oocytes developed by the growth of oogonia that migrated into the ovary before birth, accompanied by

germinal epithelial cells that become the granulosa cells. No additional primordial follicles develop after birth, when there are about 200,000 per ovary in a female human being. Their number is reduced throughout childhood and the reproductive years, and so very few, if any, remain at menopause.

During an ovarian cycle, which lasts about 28 days in human beings, certain primordial follicles enlarge and move deeper into the cortex to become **primary follicles.** The primary oocyte enlarges as food accumulates in the egg; the granulosa cells first become cuboidal or columnar in shape, and then they multiply to form several layers. Cells derived from the surrounding connective tissue envelop the granulosa cells and thus form a covering, or **theca,** for the follicle.

As development continues, the follicle becomes oval in shape, and the primary oocyte is located near one pole and is surrounded by a light-staining region known as the **zona pellucida.** A space, or **antrum,** develops amid the granulosa cells and becomes filled with a follicular liquid. This stage is a **secondary,** or **vesicular, follicle.**

Follicular growth continues until almost the middle of the ovarian cycle (from 10 to 14 days after the onset of menstruation in human beings). A **mature follicle,** which will not be seen in most slides, is similar to a vesicular follicle but much larger, extending from the medulla and causing a conspicuous outward bulge on the ovary surface. Normally only one follicle matures per ovarian cycle in human beings, but many mature per cycle in mammals that give birth to litters of offspring.

A rapid accumulation of follicular liquid and increased pressure within the antrum lead to **ovulation.** The follicle and ovary surface rupture, and the egg, with a few surrounding granulosa cells, is discharged into the coelom from which it is carried into the uterine tube by cilliary currents. In the 30-odd years of a woman's reproductive life, about 400 follicles will mature and ovulate. It follows that most of the 400,000 primordial follicles present at birth will degenerate. This degeneration occurs throughout childhood and the reproductive years and at all stages of follicle development. You may be able to observe atritic follicles, in which the eggs are beginning to shrink and the follicle walls become invaded by connective tissue and blood vessels.

A few hours before ovulation, the diploid primary oocyte undergoes its first meiotic division, which results in one **secondary oocyte** that contains most of the stored food and a very small **first polar body.** Polar bodies are seldom seen in sections of an ovary. The second meiotic division, which will produce a haploid

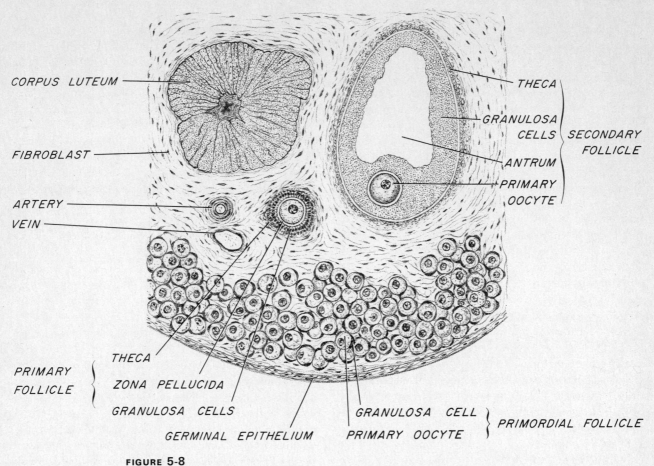

CORPUS LUTEUM

FIBROBLAST

ARTERY

VEIN

THECA

GRANULOSA CELLS } SECONDARY FOLLICLE

ANTRUM

PRIMARY OOCYTE

PRIMARY FOLLICLE {

THECA

ZONA PELLUCIDA

GRANULOSA CELLS

GERMINAL EPITHELIUM

GRANULOSA CELL } PRIMORDIAL FOLLICLE

PRIMARY OOCYTE

FIGURE 5-8
Diagrammatic microscopic section of a segment of the cortex of a cat ovary.

ootid and the **second polar body,** may be initiated, but it is not completed unless the egg is penetrated by the head of a spermotozoon—that is, unless it is fertilized. Notice that only a single mature ootid develops from one primary oocyte. This is an adaptation that conserves all of the food in one cell and provides the fertilized egg with sufficient reserves until it becomes established in the uterus and a placenta develops. The polar bodies and their nuclear material undergo no further development.

After ovulation most of the collapsed follicle is transformed into a **corpus luteum,** which may be seen in some slides. The granulosa and inner thecal cells enlarge and accumulate lipids and other materials. The human corpus luteum lasts until shortly before the onset of the next menstruation and then regresses. This marks the end of an ovarian cycle. If a pregnancy ensues, the corpus luteum enlarges more and lasts well into the pregnancy.

The follicle not only protects and nourishes the developing egg, but it and the corpus luteum are en-

docrine glands. The follicle secretes **estrogen,** which is responsible for the female secondary sex characters and for the initial build-up of the uterine lining following a menstrual period. Some estrogen is produced by the corpus luteum, but its primary hormone is **progesterone.** Progesterone is needed for the final development of the uterine lining in preparation for the implantation of an embryo and for the maintenance of the uterine lining and placenta during the early stages of pregnancy. Estrogen and progesterone production are controlled by gonadotropic hormones (follicle stimulating hormone and luteinizing hormone) secreted by the anterior part of the pituitary gland (Exercise 7). Blood levels of estrogen and progesterone, in turn, affect the release of gonadotropic releasing factors from the hypothalamus of the brain and these factors promote the release of the gonadotropic hormones. A complex and interwoven feedback mechanism between the hormones of the hypothalamus, pituitary, and ovary controls the ovarian cycle, time of ovulation, changes in the uterine lining, and pregnancy.

Nervous Coordination: Sense Organs

Survival of any organism requires the harmonious interaction of its various organs and appropriate responses of the entire organism to changes in the external environment. Metabolic activities of the body are in large part regulated by the secretions of the numerous endocrine glands. (The thyroid gland, islets of Langerhans, and suprarenal gland have been mentioned in previous exercises. The hypophysis and pineal body will be considered in the study of the brain.) The circulatory system carries minute quantities of hormones secreted by a gland in one part of the body to other regions and thus achieves endocrine integration. This type of integration tends to be slow and widespread in its effects, although certain hormones have specific target organs. Feedback mechansims of various types control the activity of the endocrine glands and the blood level of their hormones.

More rapid and specific integration of the internal organs, as well as most of the animal's responses to the external environment, is mediated through the receptor organs, nervous system, and effector organs (muscles, glands, ciliated epithelia, etc.). Neurons, which are the basic components of the nervous system, may be activated by nearly any stimulus that is intense enough. A great deal of the specificity of nervous integration derives from the facts that the receptor cells of the body are attuned to specific environmental parameters (light, temperature change, mechanical displacement, chemical change) and that they are acti-

vated by very slight changes in these parameters. The receptors, in turn, activate the neurons with which they are connected. These neurons have specific interconnections with other neurons and, eventually, with appropriate effectors. Sense organs consist of specific receptor cells and the associated cells and tissue that support and protect these cells and often amplify the environmental stimulus. The eye, ear, and nose are familiar examples.

A. EYE

The eye is complex sense organ attuned to light and to changes in the visual field. The sense organ itself, the **eyeball,** is lodged within a socket of the skull, the **orbit,** and is surrounded by various accessory structures. Notice again the upper and lower **eyelids.** They are movable in terrestrial vertebrates; they protect the eyeball and help to keep its surface moist. Make an incision extending forward from the anterior corner of the eye, if you have not already done so, and notice the **nictitating membrane** (*third eyelid*). This membrane can move across the surface of the eyeball and thus helps to keep it clean.

Carefully dissect away the eyelids, noticing as you do so that the delicate membrane lining the eyelids, the **conjunctiva,** reflects over the surface of the eyeball. With a heavy pair of scissors, remove some of the surrounding bone in order to get a better view of the eyeball. Material around the eyeball should be carefully picked away. Much is simply connective tissue, but some of it has a glandular texture. These are the tear glands, or **lacrimal glands,** which continuously produce a watery secretion that flows over the surface of the eyeball, moistening and cleaning it. Tears are drained by a **nasalacrimal duct** into the nasal passages. (The duct is difficult to find, but one entrance into it, the **lacrimal punctum,** can be seen in your neighbor's eye by pulling down the lower eyelid and looking for a small pore on its edge near the most medial eyelash. A similar pore is located on the upper eyelid.) Narrow, bandlike muscles will be seen extending from deep within the orbit to attach onto the surface of the eyeball. There are seven of them, but you may not find all of them. Collectively, they are the **extrinsic ocular muscles,** and they control the movements of the eyeball as a whole. As you pick them away, you will find the **optic nerve** emerging from the skull and entering the deep surface of the eyeball. Sever it and remove the eyeball.

Dissect the eyeball of your specimen or a larger eye from a sheep or cow. If you study a larger eye, first remove the fat and extrinsic ocular muscles and find the optic nerve, which attaches to the medial side of the eyeball slightly anterior and ventral to its center. This will help you orient the specimen. The eyeball is composed of three layers. The outermost, a dense fibrous layer, is the **fibrous tunic.** The posterolateral part of the fibrous tunic is an opaque **sclera,** but at the anterior surface of the eyeball this tunic is a transparent **cornea.** Although the cornea is rather cloudy in preserved specimens, one can usually look through it and see the pigmented **iris,** with a circular opening, the **pupil,** in its center.

Delicate structures within the eyeball will be supported better and can be seen more clearly if the eye is submerged in a dish of water while being dissected. With fine scissors cut a tangential slice from the surface of the eyeball (Fig. 6-1). The slice should be large enough to extend through the cornea, iris, and the back of the eyeball—but do not try to cut through the hard lens. The dark, pigmented **choroid,** which lies beneath the sclera, is part of the second eyeball layer, the **vascular tunic.** Its blood vessels help to nourish the eyeball. Its pigment, which is partly derived from an embryonic layer of the retina, absorbs extraneous light in the manner of the black paint inside a camera. Notice that the iris is an extension of the vascular tunic in front of the **lens.** Circular and radial muscle fibers within the iris control the diameter of the pupil, and thus the amount of light passing through the lens.

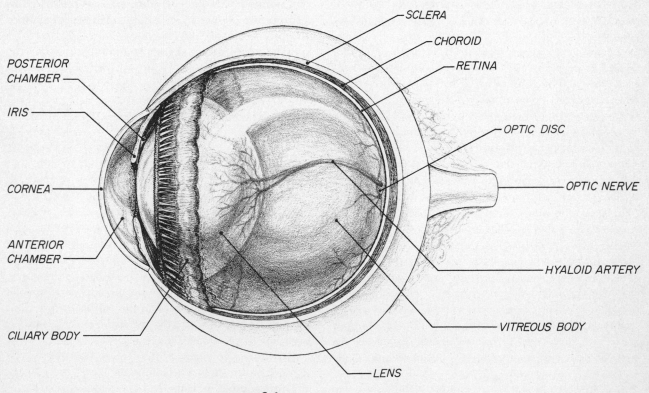

FIGURE 6-1
Dissection of the eyeball of a fetal pig.

Notice that some of the vascular tunic adheres to the periphery of the lens. Under low magnification, it will be seen that this portion has a pleated appearance. All of this is the **ciliary body.** Microscopic **zonule fibers** pass from it to the lens. Most of the refraction of light rays occurs as they pass through the cornea, which has a much higher refractive index than air, but the sharp focusing of the image is accomplished by the ciliary body acting on the lens. Intraocular fluid exerts an outward pressure on the wall of the eyeball and this produces tension on the elastic lens. Zonule fibers extend from the lens to the ciliary body, which tends to move away from the lens as intraocular pressures increase. When the eye is at rest, therefore, the lens is somewhat flattened, and distant objects are in focus. Accommodation for near objects is accomplished by the contraction of muscle fibers within the ciliary body, which in effect brings the ciliary body closer to the lens, releases the tension on the zonule fibers, and permits the lens to bulge slightly.

A sharp image falls on the **retina,** the whitish, third layer of the eyeball situated beneath the choroid. **Rods** and **cones,** the photoreceptive cells, lie on the choroid side of the retina, so that light must pass through the retina to stimulate them. Rods are adapted for vision in weak light; cones, in brighter light. Cones also mediate color vision.

The space between retina and lens is filled with a viscous **vitreous body** that helps hold the lens and retina in place. During fetal life, a small **hyaloid artery** enters the eyeball with the optic nerve, crosses the vitreous body, and supplies the developing lens. Look for it. It disappears about the time of birth. Its continued presence would cast a shadow on the retina. Remove the vitreous body to see the retina more clearly. Rods and cones are absent from the round **optic disc** where the optic nerve leaves the retina; so this is a blind spot. Examine the choroid behind the retina. In the eyes of some mammals, including the sheep and cow, the upper part of the choroid is modified and has an iridescent sheen. This is the **tapetum lucidum.** It reflects some of the light passing through the retina back onto the rods and cones and thus enables the animals to see better in dim light.

The space between the lens and the iris, the **posterior chamber,** and that between the iris and the cornea, the **anterior chamber,** are filled with a watery **aqueous humor.** Aqueous humor is secreted by the ciliary body and drained through microscopic canals at the base of the cornea. It is the aqueous humor that maintains the intraocular pressure which is antagonistic to the ciliary muscles.

The human eyeball is essentially the same as the pig's.

B. EAR

The mammalian ear has a dual function—maintaining equilibrium and detecting sound. It consists of three parts: inner ear, middle ear, and external ear. Its receptive cells lie within liquid-filled sacs and ducts of the inner ear that are embedded deep within the otic capsules of the skull (Exercise 1). Models or demonstrations of the inner ear should be examined. Its semicircular ducts, utriculus, and sacculus are related to equilibrium; hairlike cytoplasmic processes of their receptive cells are displaced by movements of the head. Receptive cells in the coiled, snaillike cochlea are activated by sounds. External and middle ears gather the sound and amplify the airborne sound waves sufficiently to set up pressure waves in the cochlear liquid.

The external ear consists of the external ear flap, the **auricle,** and an inward extending passage, the **external acoustic meatus** (Exercise 1). Cut off the auricle and cut away muscle to expose the skull ventral to the external acoustic meatus (Fig. 6-2). Follow the meatus toward the skull, carefully cutting it away as you do so. It turns ventrally and becomes encased by bone, which can be picked away with heavy forceps. Remove enough tissue to expose the rather large ear drum, or **tympanic membrane.** You may have to cut away some of the lower jaw.

The rest of the dissection should be done with the aid of a hand lens or dissecting microscope. A little bar of bone, which is a part of the malleus, can be seen through the ear drum (Fig. 6-3). Pick away the ear drum, being careful not to dislodge the malleus, and pick away a bit of the skull just dorsal to the malleus to reveal a relatively large middle ear, or **tympanic cavity.** The slit in the anteroventral portion of the cavity is part of an **auditory tube.** Pass a probe through it; observe that it opens into the nasopharynx (Exercise 3). This passage permits the equalization of pressure on each side of the tympanic membrane. Three auditory ossicles (the **malleus, incus,** and **stapes**) are situated in the dorsal portion of the cavity. The **chorda tympani,** a branch of the facial nerve supplying taste buds and certain salivary glands, crosses the malleus. The innermost ossicle, the stapes, fits into a small hole (the oval window, or **fenestra vestibuli**) in the plate of bone forming the medial wall of the cavity. The fenestra vestibuli leads to the cochlea of the inner ear. The pressure of the sound waves is amplified by the ossicular lever system and by concentration of most of the energy that impinges on the rather large tympanic membrane at the relatively small fenestra vestibuli. Slightly ventral and posterior to the fenestra vestibuli, you will see a round opening, the round window, or

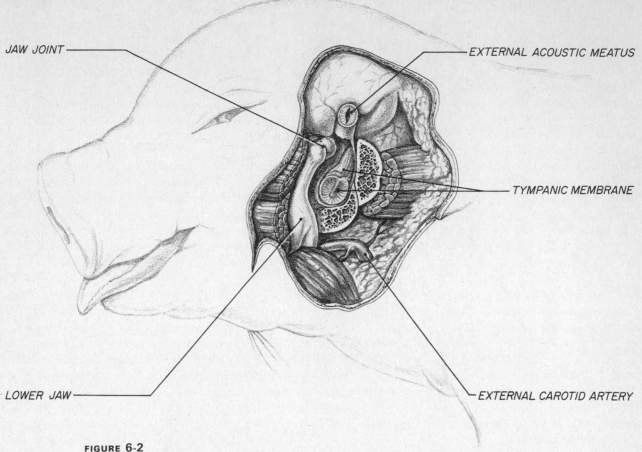

JAW JOINT

EXTERNAL ACOUSTIC MEATUS

TYMPANIC MEMBRANE

LOWER JAW

EXTERNAL CAROTID ARTERY

FIGURE 6-2
Dissection of the head of a fetal pig to show location of tympanic membrane. Bone lying lateral to much of the external acoustic meatus has been cut away.

fenestra cochlea, through which pressure waves are released from the cochlea back into the tympanic cavity. Carefully remove the ossicles, let them dry, and try to distinguish them. A glimpse of the **cochlea** may be had by carefully removing some of the bone forming the medial wall of the tympanic cavity.

The human middle ear is essentially the same.

C. NOSE

Parts of the nose were studied with the respiratory system on demonstration sagittal sections of the head (Fig. 3-2). Dissect it on your specimen at this time. Open the left **nasal cavity** by making a longitudinal section through the snout parallel to the sagittal plane. The section should pass through the **external nostril** (*naris*). Do not extend the cut into the cranial cavity. Pick away the folds of tissue in the left nasal cavity until you reach the cartilaginous **nasal septum** that separates left and right nasal cavities. As you expose the nasal septum, look for the **vomeronasal organ** near the base of the anterior part of the septum (Fig. 6-4). It is a small, tubular sac that connects with the roof of the mouth through a small, slit-shaped **incisive foramen.**

The organ is supplied by a special branch of the olfactory nerve and presumably enables the animal to smell food entering the mouth. Although this organ starts to develop in the human fetus, it is lost by birth.

Carefully cut away the nasal septum and expose the contents of the right nasal cavity. The cavity is largely filled with folds of tissue, the **nasal conchae,** whose pattern is shown in Figure 6-4. The air passages between them are the **nasal meatuses.** The meatuses converge posteriorly and enter the nasopharynx by way of the **internal nostril** (*choana*). The conchae increase the surface area available for olfaction and conditioning the respiratory air. If your specimen is well injected, you can see the extensive vascular network in the mucous membrane of the conchae. As inspired air crosses the conchae, it is warmed and moistened and dirt is entrapped in a sheet of mucus secreted by the membrane. The ethmoid conchae lie just anterior to the rostral cranial cavity, which contains the olfactory lobe of the brain, and they, in particular, increase the olfactory surface. If you dissect them carefully, you may see branches of the olfactory nerve that supplies them. Correlated with our reduced sense of smell, human conchae are neither as large nor as complexly folded.

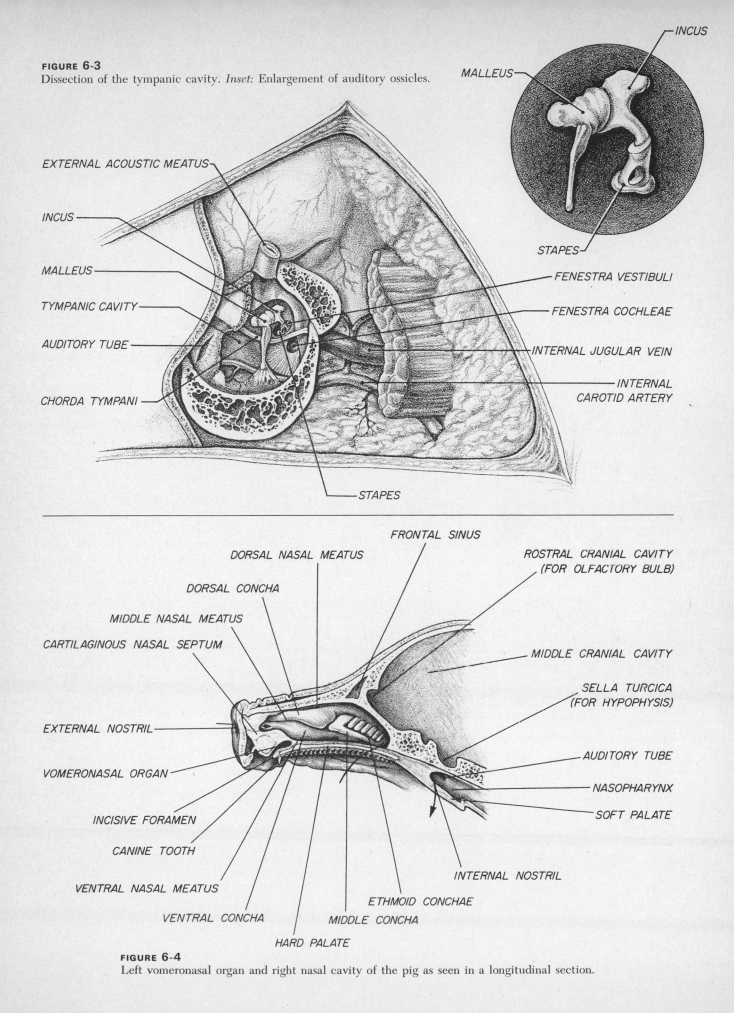

FIGURE 6-3
Dissection of the tympanic cavity. *Inset:* Enlargement of auditory ossicles.

INCUS

MALLEUS

STAPES

EXTERNAL ACOUSTIC MEATUS

INCUS

MALLEUS

TYMPANIC CAVITY

AUDITORY TUBE

CHORDA TYMPANI

FENESTRA VESTIBULI

FENESTRA COCHLEAE

INTERNAL JUGULAR VEIN

INTERNAL CAROTID ARTERY

STAPES

FRONTAL SINUS

DORSAL NASAL MEATUS

DORSAL CONCHA

MIDDLE NASAL MEATUS

CARTILAGINOUS NASAL SEPTUM

ROSTRAL CRANIAL CAVITY (FOR OLFACTORY BULB)

MIDDLE CRANIAL CAVITY

SELLA TURCICA (FOR HYPOPHYSIS)

EXTERNAL NOSTRIL

VOMERONASAL ORGAN

INCISIVE FORAMEN

CANINE TOOTH

VENTRAL NASAL MEATUS

VENTRAL CONCHA

HARD PALATE

MIDDLE CONCHA

ETHMOID CONCHAE

INTERNAL NOSTRIL

AUDITORY TUBE

NASOPHARYNX

SOFT PALATE

FIGURE 6-4
Left vomeronasal organ and right nasal cavity of the pig as seen in a longitudinal section.

Nervous Coordination: Nervous System

Grossly, the nervous system can be divided into the central nervous system, which consists of the brain and spinal cord, and a peripheral nervous system, which is composed of the cranial, spinal, and autonomic nerves. Most of the peripheral nerves are mixed, being made up of hundreds of processes of sensory and motor neurons that carry impulses from receptors to the central nervous system and from the central nervous system to the effectors. The interconnections between sensory and motor neurons and the integrative activities of the nervous system take place within the central nervous system.

A. SPINAL CORD AND SPINAL NERVES

1. Dissection of Spinal Cord and Spinal Nerves

The spinal cord and the origin of the spinal nerves can be exposed easily in the fetal pig because the vertebral column, in which they lie, is not completely ossified. Skin the back of your specimen in the thoracic region, and cut away enough back muscles to expose about 8 centimeters of the vertebral column. Completely expose the spinous processes and vertebral arches of the vertebrae (Exercise 1). As this is done, threadlike **dorsal rami** of spinal nerves (Fig. 7-1) may be seen extending into the muscles of the back. It is difficult to save them.

Using scissors, cut off the tops of the vertebral arches to expose the **vertebral canal,** in which the spinal cord lies. The **spinal cord** and **spinal nerves** are surrounded by connective-tissue sheaths known as meninges, the most conspicuous of which is the tough **dura mater.** Notice that each nerve bears an enlargement, the **spinal ganglion** (dorsal root ganglion), which is situated more or less in the **intervertebral foramen** through which the spinal nerve leaves the vertebral canal. The ganglia will be more conspicuous if you break away the vertical bar of bone between successive intervertebral foramina.

As can be seen in Figure 7-1, the spinal nerves that arise from the cord are formed by the union of their dorsal and ventral roots. To see the roots, it is necessary to slit the dura mater with a fine pair of scissors and carefully peel it to the side. It will be seen that each **dorsal root** in reality consists of a half dozen or more tiny **dorsal rootlets,** which come together at the spinal ganglion. Similarly, the **ventral root** consists of many **ventral rootlets.**

Just beyond the point at which the roots unite, a spinal nerve divides into branches, or rami (Fig. 7-1). A **dorsal ramus** supplies the muscles and skin of the back, a **ventral ramus** those of the flanks, and one or more communicating rami carry sympathetic fibers to the sympathetic cord. It is doubtful that you will be able to see the communicating rami.

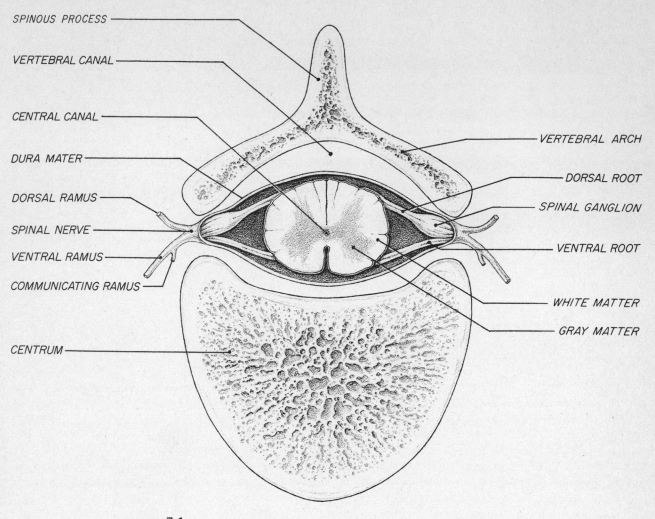

SPINOUS PROCESS

VERTEBRAL CANAL

CENTRAL CANAL

DURA MATER

DORSAL RAMUS

SPINAL NERVE

VENTRAL RAMUS

COMMUNICATING RAMUS

CENTRUM

VERTEBRAL ARCH

DORSAL ROOT

SPINAL GANGLION

VENTRAL ROOT

WHITE MATTER

GRAY MATTER

FIGURE 7-1
Diagrammatic cross section through a thoracic vertebra and the spinal cord.

2. Microscopic Section of the Spinal Cord and Spinal Nerves

Further details of the spinal cord and nerves can be seen by examining microscopic slides of these structures from a frog or cat (Fig. 7-2). The ventral surface of the **spinal cord** can be recognized by a conspicuous **ventral fissure.** As in all chordates, the spinal cord of vertebrates is hollow and contains a small **central canal** lined by a nonnervous **ependymal epithelium.** A small amount of lymphlike cerebrospinal fluid circulates in this space. The rest of the cord and the spinal nerves are composed primarily of elongated nerve cells, or **neurons.** They are distributed in such a way within the cord that one can recognize a centrally located, somewhat butterfly-shaped **gray matter** and a peripheral **white matter.** These terms derive from the color of these areas in fresh preparation; with certain stains the

gray matter may appear lighter than the white matter on the slides.

The gray matter contains the cell bodies of neurons and fibrous processes that are not covered by a fatty myelin sheath. The **cell bodies of the motor neurons** form a conspicuous group in the ventrolateral part of the gray matter. They are large, triangular or diamond-shaped cells with a conspicuous, clear nucleus containing a prominent, eccentric nucleolus. Nerve impulses from other neurons are received directly by these cell bodies or by short processes known as **dendrites** that attach to most of the angles of the cell bodies (Fig. 7-7). Impulses then travel along long processes, the **axons,** one of which leaves each cell body, passes through the **ventral root** of a **spinal nerve,** and is distributed to the skeletal muscles of the body.

The white matter consists of myelinated axons of neurons that carry impulses from the cord to centers in the brain or from the brain down to the motor neurons. These are seen in cross section. The few small nuclei that you may see among them belong to cells that ensheath the axons and deposit the myelin, to connective tissue, or to cells in the walls of blood vessels.

Sensory neurons from receptor cells enter the dorsal root of a spinal nerve because their cell bodies are located in the **spinal ganglion.** From here impulses travel along sensory axons into the spinal cord. Some ascend in the white matter to the brain. Others enter the dorsal part of the gray matter where they may terminate and be relayed elsewhere by **internuncial neurons,** whose cell bodies are difficult to distinguish, or they may continue through the gray matter to terminate on the cell bodies of the motor neurons. The termination of a sensory neuron either directly on a motor neuron or indirectly through an internuncial neuron is responsible for simple **spinal reflexes.**

In some slides, a **sympathetic ganglion** may be seen attached to the ventral surface of a spinal nerve a short distance distal to the union of the roots of the nerve (see the next section).

3. The Sympathetic Cord and Autonomic Nervous System

Turn your pig on its back, and pull open the left side of the thorax as wide as you can. Pull the heart and lungs ventrally, and push the thoracic wall dorsally. Slightly dorsal and lateral to the aorta and left azygos vein is a slender nerve band extending longitudinally against the thoracic wall. This is the **sympathetic cord,** and the segmental enlargements on it are **sympathetic ganglia.** The communicating rami of the spinal nerves connect with these ganglia. Another cord lies on the right side of the body. The cervical extension of a sympathetic cord was seen earlier lying deep to the vagus nerve between the common carotid artery and internal jugular vein (Exercise 4).

The sympathetic cords can be traced caudally into the abdomen, where they lie dorsal to the aorta or caudal vena cava. At about the level of the diaphragm, several threadlike **splanchnic nerves** extend from each sympathetic cord to **collateral ganglia,** which surround the bases of the coeliac and cranial mesenteric arteries. Most of these structures were picked away when the arteries were dissected (Exercise 4). Delicate sympathetic fibers, which cannot be seen grossly, lead from the collateral ganglia along the arteries to the abdominal viscera. The head and neck receive sympathetic fibers by means of the extension of the sympathetic cords into the neck.

The sympathetic cords are a part of the autonomic nervous system. This system consists of special motor neurons that supply most of the glands of the body and the muscles of the heart, blood vessels, urinary bladder, and gut wall. A unique feature of the system is that there is always a peripheral relay; that is, the motor neuron that leaves the central nervous system synapses with a second motor neuron whose cell bodies lie in a peripheral ganglion, and the second neuron

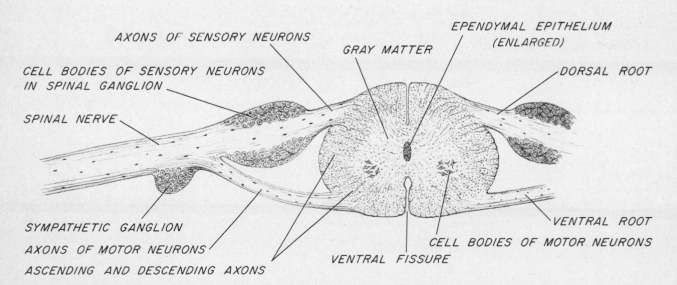

FIGURE 7-2
Microscopic transverse section of the spinal cord and spinal nerve of a frog. (After a drawing by P. Anne Smith.)

proceeds to the effector organ. The autonomic nervous system consists of a sympathetic part and a parasympathetic one. Most organs supplied by it receive nerve fibers from both parts: the two types of fibers carry impulses that have antagonistic effects on the organs.

Sympathetic fibers leave the central nervous system through the thoracic nerves and certain lumbar spinal nerves; they enter the sympathetic cord, and from it are distributed to the organs they supply. The peripheral relay between the sympathetic fibers occurs in the sympathetic ganglia on the sympathetic cord or in the collateral ganglia. Cell bodies of the second sympathetic motor neurons can be seen microscopically if a sympathetic ganglion is on the slide of the spinal cord and nerve (Fig. 7-2).

Parasympathetic fibers leave the central nervous system through certain cranial nerves and through sacral spinal nerves. The vagus nerve (cranial nerve X), dissected in Exercise 4, contains fibers that go to thoracic and abdominal viscera. Parasympathetic fibers to the eye and glands in the head leave through cranial nerves III, VII, and IX.

B. THE BRAIN AND CRANIAL NERVES

The structure of the mammalian brain can be better studied on a specially preserved sagittal section of a sheep brain than on the brain of the fetal pig. But if a sheep's brain is not available, the pig brain can be removed by skinning the head and carefully picking away the top and side of the skull (Fig. 7-3). The heavy membrane covering the brain should be left intact as long as possible. The following description refers to the sheep brain but can be applied to the brain of many other mammals, including the pig and human beings.

1. Meninges

The mammalian brain is surrounded by several layers of connective tissue, or **meninges.** A vascular **pia mater** closely invests the surface of the brain and follows all of its convolutions. The more delicate **arachnoid membrane** lies superficial to the pia and can usually be separated from the pia where it crosses the indentations

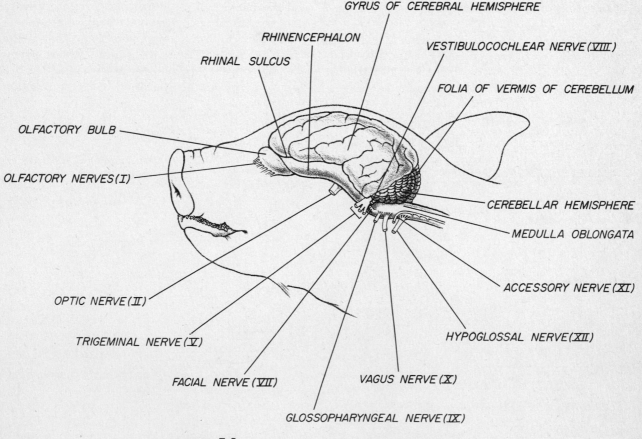

FIGURE 7-3
Lateral view of the brain and cranial nerves of a pig.

on the brain surface. In life, cerebrospinal fluid circulates between the pia and the arachnoid, forming a protective liquid cushion around the brain and helping to supply certain of its nutritive requirements. A tough, protective **dura mater** covers the other membranes. If you are studying a preserved sheep brain, the dura mater has probably been removed from your specimen, and so it should be seen on demonstration preparations.

2. Brain Regions

The different regions of the adult brain develop from five distinct swellings of the early embryonic brain, and so it is convenient to divide the adult brain into regions having the same names as the swellings (Fig. 7-4). From anterior to posterior these regions are (1) telencephalon, (2) diencephalon, (3) mesencephalon,

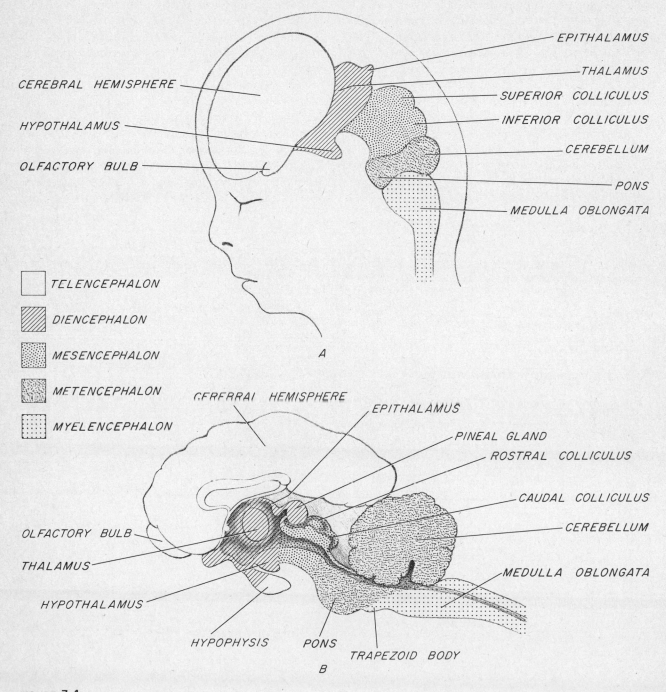

FIGURE 7.4
Brain regions as seen in (A) a lateral view of an eight-week-old human embryo and (B) a sagittal section of an adult sheep brain.

(4) metencephalon, and (5) myelencephalon.

The most conspicuous parts of the brain are the large **cerebral hemispheres** that constitute most of the **telencephalon.** They are the primary integration centers of the mammalian brain. The dorsolateral part of each hemisphere is convoluted, forming many ridges, **gyri,** that are separated by grooves, or **sulci.** Make a tangential slice across a part of a hemisphere. Notice that the surface is a gray cortex and that **white matter** lies beneath this. **Gray matter** is composed largely of the cell bodies of neurons and white matter of myelinated processes of neurons. The convolutions of the cerebral surface greatly increase the area avaliable for nerve-cell bodies.

The cerebral hemispheres of human beings are much larger and more complexly folded than those of sheep.

Sensory impulses from most parts of the body, which enter the central nervous system through the peripheral nerves, eventually reach certain of these gyri. Sensory information of different types is integrated and compared with information accumulated from past experience, and impulses are initiated in specific motor areas.

These impulses leave the central nervous system and go to appropriate effector organs through the peripheral nerves.

The conspicuous **rhinal sulcus** (Fig. 7-3 and 7-5) separates the dorsolateral, convoluted part of each cerebral hemisphere from a rather large, more ventrally situated area, the **rhinencephalon,** which is concerned with olfaction (smell). From an evolutionary viewpoint, this is one of the oldest parts of the telencephalon. In primitive fishes all of the cerebrum is concerned primarily with olfaction. An **olfactory bulb** forms the anterior part of the rhinencephalon and receives branches of the olfactory nerve entering from the nasal cavities.

The two cerebral hemispheres are connected by several bundles of white fibers that can be seen in a sagittal section. The largest and most dorsal of these commissures is the **corpus callosum** (Fig. 7-6), which connects the nonolfactory parts of the hemispheres. The wall of the hemisphere ventral to this is the thin **septum pellucidum.** Break through this wall and notice part of a large chamber, the **lateral ventricle,** within

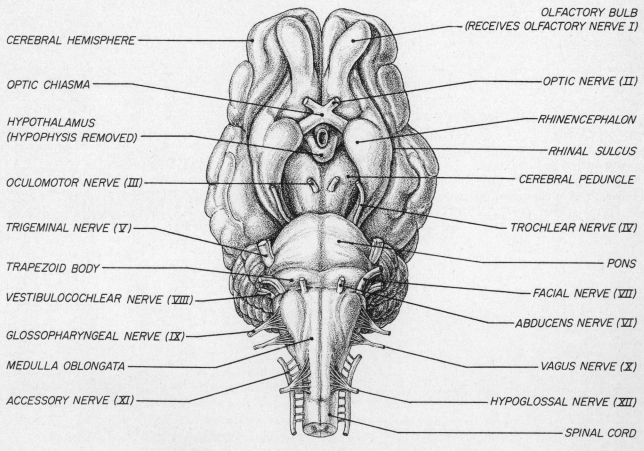

FIGURE 7-5

Ventral view of a sheep brain, and stumps of the cranial nerves.

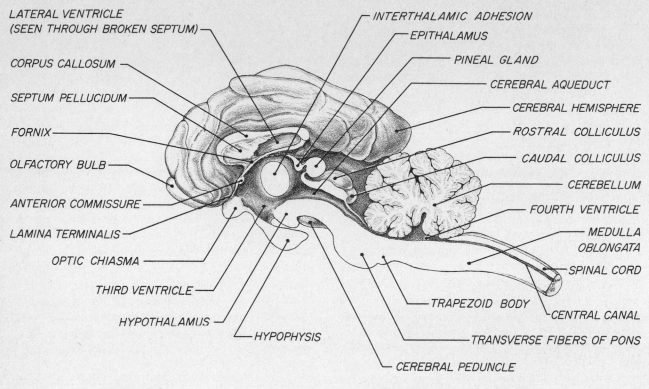

LATERAL VENTRICLE
(SEEN THROUGH BROKEN SEPTUM)

CORPUS CALLOSUM

SEPTUM PELLUCIDUM

FORNIX

OLFACTORY BULB

ANTERIOR COMMISSURE

LAMINA TERMINALIS

OPTIC CHIASMA

THIRD VENTRICLE

HYPOTHALAMUS

HYPOPHYSIS

INTERTHALAMIC ADHESION

EPITHALAMUS

PINEAL GLAND

CEREBRAL AQUEDUCT

CEREBRAL HEMISPHERE

ROSTRAL COLLICULUS

CAUDAL COLLICULUS

CEREBELLUM

FOURTH VENTRICLE

MEDULLA OBLONGATA

SPINAL CORD

CENTRAL CANAL

TRANSVERSE FIBERS OF PONS

TRAPEZOID BODY

CEREBRAL PEDUNCLE

FIGURE 7-6
Sagittal section of a sheep brain. See Figure 3-3 for sagittal section of the pig brain.

the hemisphere. There is one in each hemisphere, and they constitute the first and second ventricles. Another band of white fibers, the **fornix,** lies ventral to the septum pellucidum. It curves anteriorly and ventrally toward the hypothalamus but soon disappears from the sagittal plane. The fornix is part of a pathway that interconnects olfactory centers in the telencephalon with the hypothalamus and thalamus. It also seems to be involved in emotional behavior. Near the point of its disappearance from the sagittal plane is a small round bundle of fibers, the **anterior commissure,** which connects the major olfactory parts of the two hemispheres. A thin vertical partition extends from the anterior commissure to the crossing of the optic nerves, the **optic chiasma,** on the ventral surface of the brain. This is the **lamina terminalis,** the most anterior part of the embryonic brain in the midline. The cerebral hemispheres develop as anterolateral expansions from this region.

The cerebral hemispheres of lower vertebrates are relatively small and are situated entirely anterior to the other regions of the brain. As they assume additional functions in the course of evolution, they enlarge and grow caudally over the dorsal surface of much of the brain. The overgrowth that has occurred is evident in the sagittal section. The major brain region posterior

to the telencephalon, which is largely covered by the cerebral hemispheres, is the **diencephalon** (Fig. 7-4). It extends from the fornix, anterior commissure, and lamina terminalis posteriorly to include a conspicuous bump on the brain surface, the **pineal gland** (Fig. 7-6). The diencephalon contains a narrow chamber, the **third ventricle,** whose outlines can be recognized by its shiny lining of ependymal epithelium. (Further dissection may be necessary to expose the ventricle.) This ventricle is crossed by a large circular mass of nervous tissue, the **interthalamic adhesion.** Inconspicuous **interventricular foramina** connect the two lateral ventricles with the third. Vascular tissue forms much of the roof of the diencephalon, and tufts from it dip into the ventricle. These tufts form one of the **chorioid plexuses**—vascular networks that secrete cerebrospinal fluid into the ventricles.

The diencephalon can be divided into the **epithalamus,** lying dorsal to the third ventricle; the **hypothalamus,** ventral to it; and the halves of the **thalamus** on each lateral side of the ventricle. The epithalamus consists of the vascular roof of the ventricle, the pineal gland, and adjacent nervous tissue related to olfaction. The pineal gland is an endocrine gland that seems to have an influence on gonadial development. In some species, secretions of the gland inhibit gonadial development. Exposure to prolonged light removes this inhibition and promotes gonadial development.

The hypothalamus is the conspicuous oval area that can be seen on the ventral surface of the brain posterior to the optic chiasma. The **hypophysis,** or pituitary gland, attaches to this area; part of this gland develops embryonically as an outgrowth from the diencephalon. The hypothalamus is concerned with the integration of many autonomic functions: sleep, body temperature, water balance, appetite, and carbohydrate and fat metabolism. It exerts its influence by means of motor pathways that go out from it and by producing hormones that affect the activity of the hypophysis. The hypophysis produces a wide variety of hormones that affect many organs, including many other endocrine glands. Its hormones include one that promotes growth; others that act on the thyroid, adrenal cortex, and gonads (follicle stimulating hormone and luteinizing hormone, Exercise 5), one that promotes the development of mammary glands during pregnancy and the secretion of milk; and one that regulates water reabsorption in the kidneys. The mammalian hypophysis also produces a hormone that, when injected into a lower vertebrate, stimulates the dis-

persal of pigment. The function of this hormone in mammals is not clear.

The thalamus is very large in mammals. It includes the interthalamic adhesion and all of the lateral wall of the diencephalon. Its lateral part, which is covered by the cerebral hemisphere, can be exposed by carefully pulling the posterior end of the cerebral hemisphere laterally. (Do not tear the brain.) The thalamus, like the cerebrum, has increased in importance with evolution. All sensory information, apart from olfaction, entering the spinal cord or brain on primary sensory neurons proceeds to the thalamus on internuncial neurons (Fig. 7-7). On the way, most of these decussate, or cross to the opposite side of the body. Sensory information is sorted in the thalamus, and some or all of it is relayed on other internuncial neurons to appropriate sensory areas of the cerebral cortex. Some involuntary motor activity initiated in the cortex is also relayed in the thalamus on its way to lower motor centers. But the thalamus is more than a switchboard. Although many of its functions are as yet undiscovered, there is much evidence that it interacts with the cerebral cortex in many of the higher mental processes.

The third major region of the brain, the **mesenceph-**

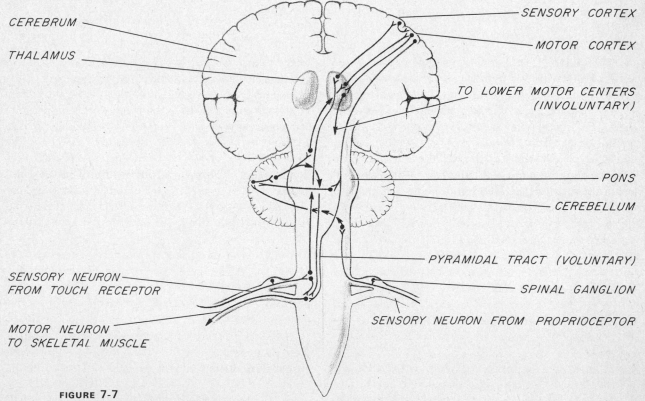

FIGURE 7-7
A wiring diagram in a ventral view of the nervous system showing the interrelations of neurons that form some of the major sensory and motor pathways. Each dark, round spot is the cell body of a neuron, and lines extending from them are axons. Impulses travel from the cell bodies to the ends of the axons.

alon, lies between the diencephalon and the large cerebellum (Fig. 7-4). It, too, is largely covered by the cerebrum and so can best be seen in a sagittal view. Dorsally, it bears two pairs of round swellings; each of the larger and more anterior is the **rostral colliculus** (superior colliculus) and each of the others is a **caudal colliculus** (inferior colliculus) (Fig. 7-6). The rostral colliculi, also known as the optic lobes, are the major integration center in the brain of fishes and amphibians, but in mammals most of their integrative function has been transferred to the cerebrum. However, certain fibers of the optic nerve still end in the rostral colliculi, and they retain their function as centers for certain eye reflexes, including the pupillary reflex, accommodation, and eyeball movements. Certain auditory reflexes occur in the caudal colliculi. The **trochlear nerve** to one of the extrinsic ocular muscles leaves the brain just posterior to the caudal colliculus. Ventrolaterally, the mesencephalon bears a pair of large fiber tracts, the **cerebral peduncles** (Fig. 7-5). The most important voluntary motor pathway, known as the **pyramidal system,** passes through these peduncles on its way from motor cell bodies in the cerebral cortex to the cell bodies of motor neurons of certain cranial nerves and the spinal nerves (Fig. 7-7). Like the ascending sensory fibers, most of the descending pyramidal fibers decussate before reaching the motor neurons of cranial and spinal nerves. **Oculomotor nerves** to most extrinsic ocular muscles attach to the peduncles. The **cerebral aqueduct** runs through the center of the mesencephalon and connects the third ventricle with the large **fourth ventricle** located in the posterior parts of the brain (Fig. 7-6).

The large **cerebellum** and the ventral part of the brain to which it attaches constitute the fourth major brain region, the **metencephalon** (Fig. 7-4). The cerebellum resembles the cerebrum in being composed of many folds, the **folia,** and in having a gray cortex. It can be divided into several parts, the most conspicuous of which are a median **vermis** and lateral **hemispheres** (Fig. 7-3). The primary sensory inflow of the cerebellum is from the part of the ear concerned with equilibrium and from the proprioceptive organs in muscle that indicate their current state of contraction. It also receives copies, so to speak, of motor directives from the cerebrum (Fig. 7-7). These terminate in the ventral part of the metencephalon, the **pons.** Neurons receiving these impulses decussate in the **transverse fibers of the pons** and lead to the cerebellar cortex. The cerebellum is therefore a crucial center for muscular coordination, monitoring the orientation and motor activity of the body, and initiating corrective impulses that either go back to the cerebrum by way of the thalamus or go directly to the cell bodies of the motor neurons. Several cranial nerves attach to the pons and their stumps can be found on good specimens (see Fig. 7-3 and 7-5): trigeminal, abducens, facial, and vestibulocochlear.

The fifth and final region of the brain is the **myelencephalon** (Fig. 7-4). It consists of the **medulla oblongata,** which merges with the spinal cord. At the front of the underside of the medulla, just posterior to the pons, there is a small transverse band of fibers, the **trapezoid body** (Fig. 7-5). This is an acoustic commissure. The fourth ventricle continues from the metencephalon into the medulla and has a very thin, vascular roof that forms another chorioid plexus. Microscopic perforations in the roof permit cerebrospinal fluid to escape from this ventricle and circulate between the pia mater and arachnoid membrane. The fluid is ultimately reabsorbed from certain parts of the meninges into the venous system. Many visceral activities are regulated by centers in the medulla: rate of heart beat, blood pressure, breathing movements, salivation, swallowing. Stumps of the remaining cranial nerves can be found attached to its surface on good specimens: glossopharyngeal, vagus, accessory, and hypoglossal.

3. Cranial Nerves

The stumps of many of the cranial nerves have been found in your study of the brain. All of these and others are shown in Figures 7-3 and 7-5. Mammals have twelve pairs, and each is given a conventional number, according to its position in human beings, as well as a name:

I. Olfactory. This consists of groups of processes of olfactory cells entering the olfactory bulb from the nose. These processes are usually torn off when the brain is removed.

II. Optic. The sensory nerve from the retina, whose fibers terminate in the thalamus and rostral colliculus.

III. Oculomotor. Primarily a motor nerve to most of the extrinsic ocular muscles, it also carries parasympathetic motor fibers to the ciliary body and pupil. Like most motor nerves, it carries a few sensory proprioceptive fibers returning from the muscles.

IV. Trochlear. A motor nerve to one of the extrinsic ocular muscles.

V. Trigeminal. A mixed nerve. It is the primary

cutaneous sensory nerve of the head, but it also carries motor fibers to the jaw muscles.

VI. Abducens.   The last of the motor nerves to the extrinsic ocular muscles; supplies one muscle.

VII. Facial.   A mixed nerve. It is the motor nerve of facial muscles (Exercise 2) and it supplies parasympathetic fibers to salivary and tear glands. Its sensory fibers come in from taste buds on the anterior two-thirds of the tongue.

VIII. Vestibulocochlear, or acoustic.   The sensory nerve from the ear.

IX. Glossopharyngeal.   A mixed nerve. It is the motor nerve to pharyngeal muscles, and it carries parasympathetic fibers to salivary glands; its sensory fibers come in from tastebuds on the posterior third of the tongue and from part of the pharynx lining.

X. Vagus.   A mixed nerve. Its motor fibers supply part of the pharynx, larynx, heart, stomach, and intestinal region. The motor fibers in this nerve that go to the heart and gut and those in the third, seventh, and ninth nerves going to the ciliary body, iris, tear glands, and salivary glands belong to the parasympathetic part of the autonomic nervous system. Its sensory fibers return impulses from many internal organs: larynx, lungs, heart, stomach.

XI. Accessory, or spinal accessory.   A motor nerve to certain neck and shoulder muscles that evolved from gill arch muscles of a fish (trapezius, sternomastoid, cleidomastoid, omohyoid—Exercise 2). It also contains the usual proprioceptive fibers.

XII. Hypoglossal.   A motor nerve to the muscles of the tongue.

Glossary of
Vertebrate Anatomical Terms

The following glossary consists of a basic vocabulary of anatomical terms that students will encounter in *Dissection of the Fetal Pig, Dissection of the Rat, Dissection of the Frog,* and many other exercises dealing with vertebrates. In addition to a definition, a derivation is given for most of the terms because an understanding of the origins of the terms will help students fix their meaning and spelling in mind. No attempt has been made to trace the history of a word, but its oldest Greek (Gr.), Latin (L.), or other (e.g., Old English) use is given. Only the nominative singular or infinitive form is listed unless the genitive (gen.), plural (pl.), or past participle (p.p.) is needed to recognize the root. Prefixes and suffixes are so indicated by an appropriately placed hyphen (e.g., *inter-,* or *-oeidos*). If a classical word has several meanings, only the one or two most appropriate to the current anatomical use of the word is included.

The origin and definition of many repetitive terms is given only under the main or first entry of the term. For example, "ligamentum arteriosum" is defined in the way that this combination of words is used, but to find the classical origins and definitions of the component parts of the term the student should also look under "ligament" and "artery." Names of individual muscles that are obviously descriptive of the muscle's location, shape, attachments, or functions have not been listed, although the component parts of less familiar ones are included. For example, the omohyoid muscle is not listed, but "omo-" and "hyoid" are. Similarly, names of blood vessels and nerves that simply state the organ supplied are omitted, but the organ itself is listed.

abducens nerve (L. *ab-,* away from + *ducere,* to lead). The fourth cranial nerve; carries motor fibers to an extrinsic ocular muscle that moves the eyeball and carries some sensory fibers from the muscle.

abduction (L. *ab-,* away from + *ducere,* p.p. *ductus,* to lead). Muscle action that pulls a body part away from a point of reference, often the midventral line of the body.

accessory nerve (Medieval L. *accessorius,* helper). The eleventh cranial nerve of higher vertebrates; carries motor fibers to certain shoulder muscles that have evolved from gill arch muscles and carries some sensory fibers from muscles.

acetabulum (L. *acetabulum,* vinegar cup, from *acetum,* vinegar). The cup-shaped socket in the pelvic girdle that receives the head of the femur.

acromion (Gr. *akros,* topmost, extreme). A process on the scapula to which the clavicle articulates in species with a well-developed clavicle.

adduction (L. *ad-,* toward + *ducere,* p.p. *ductus,* to lead). Muscle action that pulls a body part toward a point of reference, often the midventral line of the body.

adrenal gland. See *suprarenal gland.*

allantois (Gr. *allas,* gen. *allantos,* sausage). An extraembryonic membrane in reptiles, birds, and mammals that develops as an outgrowth of the urinary bladder. It accumulates waste products and serves as a gas exchange organ in reptiles and birds, and in mammals it contributes to the fetal part of the placenta.

alveoli (L. *alveolus*, small cavity). Small, thin-walled, and vascular sacs at the termination of the mammalian respiratory tree where gas exchange occurs.

ampullary gland (L. *ampulla*, small bottle). A small gland associated with the terminal end of the ductus deferens in male rats; contributes to the seminal fluid.

antebrachium (L. *ante*, before + *brachium*, upper arm). The forearm.

anterior (L. comparative of *ante*, before). A direction toward the front or belly surface of a human being; sometimes also used for the head end of a quadruped, but *cranial* is a more appropriate term except for certain structures within the head.

anterior chamber. The space within the eyeball between the cornea and iris; filled with aqueous humor.

anterior commissure. An olfactory commissure within the cerebrum that is located just anterior to the third ventricle.

antrum (Gr. *antron*, cave). An enclosed cavity within an organ, such as the antrum in a vesicular follicle.

anus (L. *anus*, anus). The caudal opening of the digestive tract on the body surface of a mammal.

aorta (Gr. *aorte*, aorta). The major artery carrying blood from the heart to the body. It is sometimes called the dorsal aorta to distinguish it from the ventral aorta of a fish, which carries blood to the gills.

aortic valve. A set of three semilunar-shaped folds at the base of the mammalian aorta that prevents a backflow of blood into the left ventricle.

aponeurosis (Gr. *aponeurosis*, to become a sinew, from *apo*, change + *neuron*, sinew, nerve). A thin, tendinous sheet.

aqueous humor (L. *aqua*, water + *humor*, liquid). Watery liquid in the anterior and posterior chambers of the eyeball; secreted by the ciliary body within the posterior chamber.

arachnoid (Gr. *arachne*, spider). One of three meninges of mammals; located between the dura mater and pia mater. It is connected to the pia mater by many connective tissue strands, which gives it a "spider web" appearance. Cerebrospinal fluid circulates in the subarachnoid space between the arachnoid and pia mater.

arbor vitae (L. *arbor*, tree + *vita*, life). The tree-shaped configuration of white matter within the cerebellum.

archinephric duct (Gr. *archi-*, primitive + *nephros*, kidney). The duct that drains the primitive kidney of a lower vertebrate; also transports sperm in a male. It becomes the ductus deferens in a male mammal.

areola (L. *areola*, small open space). A small space or area; e.g., the small, round protuberances on the surface of the pig's chorion.

arrector pili (L. *arrectus*, upright + *pilus*, hair). Small muscles in the skin of a mammal that attach to the hair follicles and elevate the hairs.

artery (L. *arteria*, artery). A vessel that carries blood away from the heart. The blood is usually rich in oxygen, but it may be low in oxygen as is that in the pulmonary arteries carrying blood from the heart to the lungs.

articular process (L. *articulare*, to divide into joints). A bony protuberance bearing a smooth surface that can move on a similar surface of an adjacent bone; specifically, one of four processes on a vertebral arch that unite successive vertebrae.

atlas (Gr. mythology *Atlas*, god supporting the heavens upon his shoulders). The first cervical vertebra that supports the skull.

atrioventricular valve (L. *atrium*, open court yard + *ventriculus*, little belly). The valve between an atrium and ventricle that prevents the backflow of blood. In a mammal, the right one has three flaps (hence, it is also called the tricuspid valve), and the left one has two flaps (sometimes called bicuspid valve or mitral valve).

atrium (L. *atrium*, open court yard). A chamber of the heart that receives blood from the body (right atrium) or lungs (left atrium).

auditory tube (L. *audire*, to hear). Passage from the middle ear, or tympanic cavity, to the pharynx. It equalizes pressure on each side of the tympanic membrane. Also known as eustachian tube.

auricle (L. *auricula*, little ear). The external ear flap; also an ear-shaped lobe on the mammalian heart atrium.

autonomic nervous system (Gr. *autos*, self + *nomos*, rule). An involuntary part of the nervous system (it "rules itself") that supplies motor fibers to glands and visceral organs; consists of sympathetic and parasympathetic divisions. Autonomic innervation involves chains of two neurons with a peripheral synapse.

axillary (L. *axilla*, armpit). Pertaining to the armpit; e.g., axillary artery.

axis (L. *axis*, hub, axle). The second cervical vertebra of a mammal; rotation of the head occurs between the atlas and the axis.

axon (Gr. *axon*, axle). A long neuron process that carries impulses away from the cell body.

azygos vein (Gr. *a-*, without + *zygon*, yolk). An unpaired vein in mammals that drains most of the intercostal spaces on both sides of the body.

basophile (Gr. *basis*, foundation + *philos*, having an affinity for). A leukocyte whose cytoplasmic granules take basic stains and appear blue; rarest of the leukocytes, it releases histamine in tissue injury.

Bowman's capsule. The thin-walled expanded proximal end of a kidney tubule that surrounds a glomerulus.

brachial (L. *brachium*, upper arm). Pertaining to the upper arm; e.g., brachial artery, coracobrachialis muscle.

brachial plexus. The network of nerves that supplies the shoulder and arm.

brachium. The upper arm.

brachium conjunctivum. An armlike neuron tract through which impulses leave the cerebellum; located near the anterior border of the cerebellum.

brachium pontis. An armlike neuron tract that carries fibers from the pons to the cerebellum.

branchiomeric muscle (Gr. *branchion*, gill + *meros*, part, segment). Muscles that have evolved from those associated with the gill arches and jaws of a fish.

broad ligament. Mesentery in a female mammal that attaches the reproductive tract to the dorsal body wall.

bronchus (Gr. *bronkhos*, windpipe). A branch of the trachea that leads to a lung.

bulbourethral gland (L. *bulbus*, bulb + *urina*, urine). A gland in male mammals near the base of the penis that contributes to the seminal fluid.

caecum or **cecum** (L. *caecus*, blind). A blind pouch at the beginning of the large intestine. It is very long in many herbivores, including the rat, and often harbors a bacterial colony that digests cellulose.

calcaneus (L. *calx*, gen. *calcis*, lime, heel). The large proximal tarsal that projects in mammals as the "heel bone."

calyx, pl. calyces (Gr. *kalyx*, cup). A cuplike compartment; e.g., the renal calyces are subdivisions of the renal pelvis.

canaliculi (L. *canaliculus*, little canal). Minute canals in the bone matrix that allow processes of the bone cells to communicate with the spaces lodging blood vessels.

canine tooth (L. *canis*, dog). The large pointed tooth in mammals that is located caudal to the incisor teeth. In human beings, its crown resembles an incisor, but its root is much larger; absent in the rat.

capillary (L. *capillus*, hair). Minute, thin-walled blood vessels located between small arteries and veins through whose walls exchanges between the blood and interstitial fluid occur.

cardiac (Gr. *kardia*, heart). Pertaining to the heart or the vicinity of the heart.

carotid artery (Gr. *karotides*, large artery of the neck, from *karoun*, to stupefy). One of a pair of large arteries beside the trachea that carry blood to the head. Pressure on them reduces blood flow to the brain and causes one to be stupefied.

carotid body. An enlargement at the junction of the internal and external carotid arteries that is a chemoreceptor monitoring the oxygen and carbon dioxide levels in the blood and initiating an increase in respiratory rate if needed.

carpals (Gr. *karpos*, wrist). The small bones of the wrist.

caudal (L. *cauda*, tail). Pertaining to the tail, or to a direction toward the tail in a quadruped.

caudal (**inferior**) **colliculus** (L. *colliculus*, small hill). One of a pair of small bumps on the caudodorsal surface of the mammalian mesencephalon; a center for certain acoustic reflexes.

cell body. The part of a nerve cell, or neuron, that contains the nucleus. Cell bodies are normally located in the gray matter of the central nervous system or in peripheral ganglia.

central canal (L. *centrum*, center). The cavity in the center of the spinal cord. It connects with the fourth ventricle in the medulla oblongata and is filled with cerebrospinal fluid.

centrum. The main body or center of certain organs; e.g., the disc-shaped part of a vertebra located ventral to the spinal cord.

cephalic (Gr. *kephale*, head). Pertaining to the head; e.g., brachiocephalic vein.

cerebellum (L. *cerebellum*, little brain). The dorsal part of the metencephalon. It is an important center for the motor control of muscular activity and equilibrium.

cerebral aqueduct (L. *cerebrum*, brain + *aqua*, water). A narrow passage through the mesencephalon and metencephalon that transports cerebrospinal fluid from the third to the fourth ventricle.

cerebral hemisphere. One of two hemispheres that form the cerebrum. They occupy most of the telencephalon and are the major center for integration and higher mental processes in mammals.

cerebral peduncle (L. *pedunculus*, little foot). One of a pair of conspicuous neuron tracts on the ventral surface of the mesencephalon of a mammal that carry impulses from the cerebral hemispheres to motor centers and to the pons.

cervical (L. *cervix*, neck). Pertaining to the neck; e.g., cervical vertebrae.

cervix. The neck of an organ; e.g., the cervix of the uterus.

choanae (Gr. *choane*, funnel). The internal nostrils leading from the nasal cavities to the pharynx.

chorda tympani (L. *chorda*, string + *tympanum*, drum). A branch of the facial nerve crossing the tympanic membrane and malleus on its way to innervate certain taste buds (sensory) and salivary glands (motor).

chorioid (**choroid**) **plexus** (Gr. *khorion*, afterbirth +*-oeidos*, resemblance to). Vascular tufts that project from a thin brain surface into the cerebral ventricles and secrete the cerebrospinal fluid.

chorion. The outermost extraembryonic membrane in reptiles, birds, and mammals. Together with the allantois, it forms the fetal part of the mammalian placenta.

choroid. A vascular and pigmented layer of the eyeball wall located between the sclera and the retina. It is a part of the vascular tunic.

chromatophore (Gr. *chroma*, color + *phoros* from *pherein*, to bear). A cell in the skin of lower vertebrates that contains pigment granules.

ciliary body (L. *cilium*, eyelash). A part of the vascular tunic attached to the lens. Its muscle fibers regulate accommodation and its secretory cells produce the aqueous humor.

clavicle (L. *clavicula*, small key, clavicle). The collar bone. It extends between the sternum and scapula in species in which it is well developed.

cleido- (Gr. *kleis*, gen. *kleidos*, key, clavicle). A root referring to the clavicle. Used in combination with other terms; e.g., cleidobrachialis muscle.

cloaca (L. *cloaca*, sewer). A chamber in lower vertebrates that receives the termination of the digestive tract and the urinary and genital ducts. Its division in mammals continues these passages to the body surface.

coagulating gland (L. *coagulare*, to curdle). A gland closely associated with the vesicular gland in male rats; contributes to the seminal fluid.

cochlea (Gr. *kokhlos*, land snail). Spiralled part of the inner ear of a mammal containing the auditory receptors and associated structures.

coeliac artery (Gr. *koilia*, belly). A branch of the aorta that supplies the cranial abdominal viscera including the stomach, spleen, and liver.

coelom (Gr. *koilos*, hollow). A body cavity in vertebrates and many invertebrates that develops within the mesodermal germ layer and is completely lined by a simple, squamous epithelium of mesodermal origin that reflects over the internal organs.

collagen (Gr. *kolla*, glue + L. *genere*, to beget). Minute protein fibers forming most of the intercellular material in connective tissues; the most abundant protein in the body.

colon (Gr. *kolon*, large intestine). Most of the large intestine. Depending on the species, it extends from the small intestine or the caecum to the cloaca or rectum.

columella (L. *columella*, little column). A term frequently used for the rod-shaped stapes in lower terrestrial vertebrates. See *stapes.*

commissure (L. *commissura*, from *committere*, to commit, to join together). A neuron tract that interconnects structures on the left and right sides of the central nervous system.

common bile duct (L. *bilis*, bile). The principal duct supplying bile to the small intestine. It is formed by the confluence of the hepatic ducts from the liver and, when present, the cystic duct from the gall bladder.

conchae (Gr. *konkhe*, seashell). Folds within the mammalian nasal cavities that increase the surface area.

condyle (Gr. *kondylos*, articulating prominence at a joint). A rounded articular surface.

condyloid process. The joint-bearing process of the mammalian mandible.

conjunctiva (L. *conjungere*, p.p. *conjunctus*, to join together). Thin membrane covering the surface of the cornea and reflecting onto the inner surface of the eyelids.

conus arteriosus. A chamber of the heart of lower vertebrates into which the ventricle discharges. This chamber contributes to the bases of the pulmonary trunk and aorta in mammals.

coracoid (Gr. *korax*, gen. *korakos*, crow). The bone forming the caudoventral part of the pectoral girdle in lower vertebrates.

coracoid process. A process shaped like a crow's beak on the base of the mammalian scapula median to the glenoid cavity. It is a remnant of the large coracoid region of the pectoral girdle in lower vertebrates.

cornea (L. *cornu*, horn). The transparent part of the eyeball wall through which light passes. It is a part of the fibrous tunic.

coronary vessels (L. *corona*, garland, crown). Blood vessels that encircle the heart between the atria and ventricles and supply the cardiac muscles.

coronoid process. The uppermost process of the mammalian mandible to which certain jaw-closing muscles attach.

corpus luteum (L. *corpus*, body + *luteus*, yellow). In life, a yellowish endocrine gland within the ovary that develops after ovulation from the follicle. Its principle hormone is progesterone.

cortex (L. *cortex*, bark). The outer layer of many organs; e.g., cerebral cortex, renal cortex.

costal cartilages (L. *costa*, rib). Cartilages that extend between the ribs and sternum in many vertebrates. Their flexibility permits rib movements.

cranial (Gr. *kranion*, skull, brain case). Pertaining to the brain case or to a direction toward the head end of the body in a quadruped.

cremasteric muscle (Gr. *kreas*, flesh). Muscle layer derived from the abdominal wall that surrounds the processus vaginalis. It is part of the mammalian scrotum and is particularly well developed in the rat.

crus (L. *crus*, lower leg). The shank or lower leg.

cucullaris muscle (L. *cucullus*, cap. hood). A muscle covering the craniodorsal part of the shoulder in lower vertebrates. It has evolved from certain gill arch muscles of a fish and gives rise to the mammalian trapezius and sternocleidomastoid complex of muscles.

cutaneous (L. *cutis*, skin). Pertaining to the skin; e.g., cutaneous nerve.

cystic duct (Gr. *kystis*, bladder). The duct of the gall bladder. Together with the hepatic duct from the liver, it enters the common bile duct.

deferent duct. See *ductus deferens.*

deltoid muscle (Gr. *delta*, the fourth letter of the Greek alphabet, Δ). A muscle crossing the lateral surface of the shoulder, it is shaped like the letter delta in human beings.

dendrite (Gr. *dendron*, tree). A highly branched and usually short neuron process that carries nerve impulses toward the cell body.

dermis (Gr. *derma*, skin, leather). The deeper layer of the skin composed of a dense connective tissue.

diaphragm (Gr. *dia*, across, between + *phragma*, partition). A muscular partition between the thoracic and abdominal cavities of a mammal. Its contraction is a major force in inspiration.

diencephalon (Gr. *di[a]*, across, between + *enkephalos*, brain). The second of five major brain regions. It lies between the telencephalon and the mesencephalon and includes the thalamus, epithalamus, and hypothalamus.

distal (L. *distare*, to be far away). That end of a structure most distant from its origin.

dorsal (L. *dorsum*, back). A direction toward the back surface of a quadruped.

duct (L. *ductus*, small tubular passage). A small, tubular passage usually carrying products from one organ to another or to a body surface.

ductus arteriosus. A connection in the fetal mammal between the pulmonary trunk and the aorta that permits much of the blood in the pulmonary trunk to bypass the lungs. It atrophies after birth and becomes the adult ligamentum arteriosum.

ductus deferens (L. *deferre*, to carry away). The sperm duct of a mammal that transports sperm from the epididymis to the urethra.

ductus venosus. A passage in the fetal mammal that permits much of the blood in the umbilical vein, which is returning from the placenta, to bypass the hepatic sinusoids and enter the caudal vena cava directly.

duodenum (from the Latin *intestinum duodenum digitorum; duodeni*, twelve each). The first part of the small intestine, which is about twelve fingerbreadths long in human beings.

dura mater (L. *durus*, hard + *mater*, mother). The outermost of the meninges surrounding the central nervous system. It is composed of dense connective tissue.

efferent ductules (L. *ex-* or *ef-*, out of, away from + *ferre*, to carry). Small ducts in the males of lower vertebrates that carry sperm from the testis to the cranial tubules of the kidney; comparable to the rete testis of a mammal.

eosinophile (Gr. *eos*, dawn + *philos*, having an affinity for). A leukocyte whose cytoplasmic granules stain with eosin (an acid dye) and appear red; participates in allergic reactions.

epaxial (Gr. *epi-*, upon, above + L. *axis*, axle). Pertaining to those muscles or other organs that lie beside or above the vertebral axis.

ependymal epithelium (Gr. *ependyma*, tuniclike upper garment). The epithelium that lines the spaces within the central nervous system and contributes to the chorioid plexus.

epidermis (Gr. *epi-*, upon + *derma*, skin, leather). The superficial layer of the skin. It is composed of a stratified, squamous epithelium and is usually keratinized in the upper layers.

epididymis (Gr. *epi-*, upon + *didymos*, testicle). A band-shaped group of tubules and a coiled duct that lie upon the testis of a mammal. Used for the storage of sperm, it evolved from part of the primitive kidney and archinephric duct in lower vertebrates.

epiglottis (Gr. *epi-*, upon + *glotta* or *glossa*, tongue). A trough-shaped flap supported by cartilage at the base of the tongue of a mammal and cranial to the glottis. It helps prevent food from entering the larynx.

epiphysis (Gr. *epi-*, upon + *physis*, growth). (1) The end of a long bone in a young mammal. Growth of an epiphyseal plate of cartilage between the epiphysis and the bone shaft, and its replacement by bone, results in the growth in length of the bone. (2) A stalklike outgrowth of the epithalamus in lower vertebrates whose distal end is sensitive to changes in light intensity. It is comparable to the mammalian pineal gland.

epithalamus (Gr. *epi-*, upon + *thalamos*, inner chamber). An olfactory center of the brain that lies above the thalamus and contributes to the roof of the diencephalon.

epithelium (Gr. *epi-*, upon + *thele*, nipple). A tissue composed of tightly packed cells that cover all body surfaces and line all cavities including the lumen of blood vessels and ducts. Secretory cells of glands develop embryonically from epithelial layers.

epitrichium (Gr. *epi-*, upon + *trichion*, hair). A layer of epithelium that lies upon the developing hairs in a mammalian fetus.

erythrocyte (Gr. *erythros*, red + *kytos*, hollow vessel, cell). A red blood cell. It contains hemoglobin, which combines reversibly with oxygen in the lungs and transports it to the tissues.

esophagus (Gr. *oisophagos*, gullet). That part of the digestive tract between the pharynx and the stomach.

eustachian tube. See *auditory tube.*

extension (L. *ex-*, out of or away from + *tendere*, to stretch). Muscle action that increases the angle at a joint.

external acoustic meatus (Gr. *akoustikos*, pertaining to hearing + L. *meatus*, passage). The canal that leads from the body surface to the tympanic membrane.

fallopian tube. See *uterine tube.*

fenestra (L. *fenestra*, window). A windowlike opening in an organ.

fenestra cochlea. A small opening on the median wall of the tympanic cavity of a mammal from which pressure waves that have travelled through the inner ear are released; often called the round window.

fenestra vestibuli (L. *vestibulum*, entrance chamber). A small opening on the median wall of the tympanic cavity into which the distal end of the stapes fits and sets up pressure waves in the inner ear; often called the oval window.

fibroblast (L. *fibra*, fiber + Gr. *blastos*, bud, germ). Elongated cells in connective tissue that produce collagen and other intercellular materials.

fibrous tunic (L. *fibra*, fiber + *tunica*, loose garment). Dense connective tissue forming the outermost layer of the eyeball. The transparent cornea and the opaque sclera are parts of the fibrous tunic.

fibula (L. *fibula*, buckle, pin). The slender bone on the lateral surface of the lower leg.

flexion (L. *flectere*, p.p. *flexus*, to bend). Muscle action that decreases the angle at a joint.

folia (L. *folium*, leaf). Leaf-shaped folds such as those of the cerebellar cortex of a mammal.

foramen (L. *foramen*, opening, hole). A small opening in the skull or other organ.

foramen magnum (L. *magnus*, great, large). The opening in the base of the skull through which the spinal cord passes. It is the largest skull foramen.

foramen ovale (L. *ovalis*, oval). An opening in the interatrial septum of a fetal mammal that permits much of the blood in the right atrium (chiefly blood from the caudal vena cava and placenta) to enter the left atrium and thus bypass the lungs. It closes at birth and becomes the adult fossa ovalis.

fornix (L. *fornix*, vault, arch). An arch-shaped neuron tract that leads from olfactory centers in the telencephalon to the hypothalamus; also involved in emotional behavior.

fossa (L. *fossa*, ditch). A shallow depression in an organ.

fossa ovalis (L. *ovalis*, oval). An oval-shaped depression in the median wall of the right atrium of an adult mammal that is a vestige of the fetal foramen ovale.

frontal (L. *frons*, forehead). (1) Pertaining to the forehead. (2) A plane of the body that passes through the frontal suture between the parietal and frontal bones; i.e., a median longitudinal plane passing from left to right.

gall bladder (Old English, *gealla*, bitter, sour). A small sac attached to the liver in most vertebrates that stores and concentrates bile. It is not present in rats.

ganglion (Gr. *ganglion*, swelling, tumor). An aggregation of neuron cell bodies. In vertebrates, ganglia lie peripheral to the central nervous system.

gastric (Gr. *gaster*, belly). Pertaining to the stomach; e.g., gastric artery, gastric gland.

gemelli muscles (L. *geminus*, twin-born). Two small muscles situated deeply on the lateral surface of the thigh in mammals between the obturator foramen and the proximal end of the femur.

gland (L. *glans*, acorn). A group of secretory cells of epithelial origin. Exocrine glands discharge by a duct onto the body surface or into a body cavity; endocrine glands discharge into the circulatory system.

glans (L. *glans*, acorn). The "acorn shaped" tip of the penis or clitoris.

glenoid (Gr. *glene*, cavity, socket + *-oeides*, resemblance to). The socket on the scapula that receives the head of the humerus.

glomerulus (L. *glomerulus*, little ball). A ball-like group of capillaries associated with Bowman's capsule at the proximal end of a kidney tubule.

glossopharyngeal nerve (Gr. *glossa* or *glotta*, tongue + *pharynx*, pharynx). The ninth cranial nerve, which carries motor fibers to pharyngeal muscles and parasympathetic fibers to certain salivary glands; it returns sensory fibers from taste buds on part of the tongue.

glottis (Gr. *glossa* or *glotta*, tongue). A slitlike opening at the base of the tongue into the larynx, including the space between the vocal cords.

gluteus muscle (Gr. *gloutos*, buttock). One of several buttock muscles in human beings. They are located on the caudolateral surface of the pelvic girdle in quadrupeds.

goblet cell (Middle English, *goblet*, drinking bowl with a narrow stem). A goblet-shaped, mucus-secreting cell associated with the columnar epithelium of the stomach, intestinal lining, and upper respiratory tract.

gonad (Gr. *gonos*, progeny). An ovary or testis.

gracilis muscle (L. *gracilis*, slender). A wide, but thin, muscle that is located superficially on the medial side of the thigh and adducts the leg.

gubernaculum (L. *gubernaculum*, rudder). A cord of tissue at the caudal end of the testis that guided its descent during embryonic development.

gyrus (Gr. *gyros*, circle, round). A fold of the cerebral cortex.

hallux (L. *hallux*, big toe). The first or most median digit of the foot.

hemoglobin (Gr. *haima*, blood + L. *globus*, globe). Compound in erythrocytes that combines reversibly with oxygen. It consists of an iron-containing heme plus a protein globin.

hepatic (Gr. *hepar*, liver). Pertaining to the liver.

hepatic duct. The first of the duct system that carries bile from the liver. Hepatic ducts usually join the cystic duct to form the common bile duct.

humerus (L. *humerus*, bone of the upper arm). The bone of the upper arm.

hyaloid artery (Gr. *hyalos*, glass + *oeidos*, resemblance to). The embryonic artery that passes through the vitreous body of the eyeball to supply the developing lens. It disappears at about the time of birth.

hyoid (Gr. *hyoeides*, resembles the letter upsilon). The U-shaped bone lodged in and supporting the base of the tongue in mammals.

hyoid apparatus. A group of bones and cartilages in lower terrestrial vertebrates that are associated with the base of the tongue. It includes more visceral arches than the mammalian hyoid.

hypaxial (Gr. *hypo-*, under + L. *axis*, axle). Pertaining to those muscles or other organs in the body wall that lie ventral to the vertebral axis.

hypobranchial muscles (Gr. *hypo-*, under + *branchion*, gill). Tongue muscles and many of those on the ventral surface of the neck. They have evolved from muscles that lie ventral to the gills in fishes.

hypoglossal nerve (Gr. *hypo-*, under + *glossa* or *glotta*, tongue). The twelfth cranial nerve in mammals, which carries motor fibers to the muscles of the tongue.

hypophysis (Gr. *hypo-*, under + *physis*, growth). An endocrine gland, often called the pituitary gland, that is attached by means of the infundibulum to the underside of the hypothalamus. It produces or releases a variety of hormones regulating growth, metabolism, sexual activity, and water balance.

hypothalamus (Gr. *hypo-*, under + *thalamos*, inner chamber). The floor of the diencephalon located beneath the thalamus. It is an important center for the control of visceral activity and the regulation of the hypophysis.

iliac (L. *ileum* or *ilium*, groin). Pertaining to the region of the groin or to the flank; e.g., iliac artery.

ilium. The bone forming the dorsal most part of the pelvic girdle and articulating with the sacrum.

incisor tooth (L. *incidere*, p.p. *incisus*, to cut into). One or more teeth located at the front of the mammalian jaw. Human beings have two incisors in each jaw half that are adapted for cutting or nipping; rodents have one enlarged incisor adapted for gnawing.

incus (L. *incus*, anvil). The middle of three auditory ossicles in mammals that transmit sound waves across the tympanic cavity. It is shaped like an anvil.

inferior (L. comparative of *inferus*, low). A direction toward the tail end of a human being.

infundibulum (L. *infundibulum*, funnel). A funnel-shaped region of an organ; e.g., the infundibulum at the beginning of the uterine tube.

inguinal canal (L. *inguinalis*, pertaining to the groin). A passage through the abdominal muscle layers in the groin of a male mammal through which the ductus deferens and blood vessels to and from the testis pass.

insertion of a muscle (L. *in*, in + *serere*, to sow or plant). That attachment of a muscle to a bone or other element that moves the greater distance when the muscle contracts. Usually this is the distal end of a limb muscle.

internuncial neuron (L. *inter-*, between + *nuntius*, messenger). A neuron within the central nervous system that lies between sensory and motor neurons or other internuncial neurons. Sometimes it is called an interneuron.

interstitial cell (L. *inter-*, between + *sistere*, to stand). Groups of endocrine cells between the seminiferous tubules of the testis that produce the male hormone testosterone.

interthalamic adhesion (L. *inter-*, between + Gr. *thalamos*, inner chamber). The part of the mammalian thalamus that crosses the midline and third ventricle; also called massa intermedia.

interventricular foramen (L. *inter-*, between + *ventriculus*, little belly). Foramen leading from one of the lateral ventricles of the brain to the third ventricle; also called foramen of Monro.

intervertebral foramen (L. *inter-*, between + *vertebra*, vertebra). Openings between successive vertebrae on the lateral surface of the vertebral column through which the spinal nerves pass.

intestine (L. *intestinum*, from *intestinus*, internal). The main digestive and absorptive part of the digestive tract; located between the stomach and the cloaca or anus, it is divided into small and large intestines in terrestrial vertebrates.

iris (Gr. *iris*, rainbow). The outer, pigmented part of the vascular tunic of the eyeball that surrounds the pupil. Muscles within the iris control the size of the pupil and hence the amount of light that enters the eyeball.

ischiadic (ischiatic) nerve (Gr. *ischion*, hip). Large nerve on the caudolateral surface of the hip and leg that supplies many of the muscles that flex the shank and extend the foot. It is also called the sciatic nerve.

ischium (Gr. *ischion*, hip). The bone forming the caudoventral part of the pelvic girdle.

islets of Langerhans. Patches of endocrine tissue within the pancreas that secrete hormones (insulin and glucagon) essential for carbohydrate metabolism.

jejunoileum (L. *jejunus*, empty + *ileum* or *ilium*, groin, flank). The postduodenal small intestine. It is usually found to be empty during dissections.

jugular vein (L. *jugulum*, clavicle). One of two pairs of veins in the neck that help to drain the head. They pass close to the clavicle in human beings. The larger external jugular vein lies superficially on the side of the neck; the smaller internal jugular vein, deeper beside the common carotid artery.

kidney (Middle English, *kidenei*, kidney). The organ that removes waste products, especially nitrogenous wastes of metabolism, from the blood and produces urine.

lacrimal (L. *lacrima*, tear). Pertaining to the tear or lacrimal apparatus; e.g., lacrimal bone, lacrimal gland, nasolacrimal duct.

lacunae (L. *lacuna*, pl. *lacunae*, pool). Small cavities in the bone matrix that lodge the main part of the bone cells (osteocytes).

lamella (L. *lamella*, little layer). A thin layer or plate; e.g., the layers of collagen fibers in bone matrix.

lamina terminalis (L. *lamina*, layer + *terminalis*, boundary, limit). A thin layer of nervous tissue forming the anterior boundary of the diencephalon in the midline.

laryngotracheal chamber. A chamber of the respiratory tract in lower terrestrial vertebrates into which the glottis leads and from which the lungs emerge. It is comparable to the mammalian larynx and trachea.

larynx (Gr. *larynx*, larynx). That part of the mammalian respiratory tract between the pharynx and the trachea. It contains the vocal cords and is sometimes called the "voice box."

lateral (L. *latus*, side). A direction toward the side of the body.

lens (L. *lens*, lentile). An important refractive body near the front of the eyeball that is responsible for accommodation or focussing by changes in shape (mammals) or by moving toward or away from the retina (frogs).

leukocyte (Gr. *leukos*, clear, white + *kytos*, hollow vessel, cell). Any of several different types of white blood cells.

lienic (L. *lien*, spleen). Pertaining to the spleen; e.g., lienogastric artery.

ligament (L. *ligamentum*, bond, bandage). (1) A band of dense collagenous connective tissue extending between bones. (2) A mesentery extending between certain visceral organs.

ligamentum arteriosum. A cord of tissue extending from the pulmonary trunk to the aorta in adult mammals. It is a remnant of the fetal ductus arteriosus.

linea alba (L. *linea,* line + *albus,* white). A white strand of connective tissue in the midventral abdominal wall to which the muscles of the abdominal wall attach.

liver (Old English, *lifer,* liver) The large gland in the cranial part of the abdominal cavity that secretes bile and metabolizes carbohydrates, proteins, and fats brought to it in the hepatic portal system. It also synthesizes many plasma proteins, degrades toxins, and removes damaged red blood cells.

lumbar vertebra (L. *lumbus,* loin). One of several caudal trunk vertebrae of a mammal located in the loin; i.e., the part of the back between the chest and the pelvis.

lumbosacral plexus. The network of nerves supplying the pelvis and legs.

lung (Old English, *lungen,* lung). The organ in terrestrial vertebrates in which gases are exchanged between the blood and the air. It develops embryonically as an outgrowth from the floor of the pharynx.

lymph (L. *lympha,* watery fluid). The liquid in the lymphatic vessels; differs from blood in lacking most of the plasma proteins and red blood cells. It is derived from the interstitial fluid and eventually enters the veins.

lymph heart. Pulsating part of lymphatic vessels in lower vertebrates that helps to return lymph to the veins.

lymph node. Small, oval body associated with the lymphatic vessels of higher vertebrates. Many lymphocytes are produced here, foreign particles phagocytosed, and certain immune responses initiated.

lymphocyte (Gr. *kytos,* hollow vessel, cell). A leukocyte with a large nucleus and very little cytoplasm. Lymphocytes develop in lymph nodes and other lymphoid organs, and they participate in the immune response.

malleus (L. *malleus,* small hammer). The outermost of three auditory ossicles of a mammal that transmit sound waves across the tympanic cavity. It abuts the tympanic membrane and is shaped like a small hammer.

mamillary body (L. *mamilla,* small breast). The caudal, somewhat breast-shaped part of the hypothalamus.

mammary gland (L. *mamma,* breast). The milk-secreting gland that characterizes female mammals.

mammary papilla. The nipples, or teats, of a mammary gland.

mandible (L. *mandibula,* lower jaw). The lower jaw.

mandibular cartilage. A cartilaginous core to the lower jaw of the frog. It forms the ventral half of the first visceral arch in fishes.

mandibular gland. A mammalian salivary gland usually located deep to the caudoventral angle of the mandible.

masseter muscle (Gr. *maseter,* chewer). Large mammalian muscle that helps to close the jaws. It extends from the zygomatic arch to the mandible.

mastoid process (Gr. *mastos,* breast + *-oeidos,* resemblance to). Large, rounded process on the base of the mammalian skull caudal to the external acoustic meatus to which certain neck muscles attach.

meconium (Gr. *mekoneion,* poppy juice). Bile-stained debris in the fetal digestive tract. It is discharged at birth.

median (L. *medius,* middle). A direction toward the middle of the body.

mediastinum (L. *mediastinus,* median). The area in the middle of the mammalian thorax between the two pleural cavities. It contains the aorta, esophagus, pericardial cavity and heart, thymus, and vena cava.

medulla (L. *medulla,* marrow). The core of certain organs; e.g., the renal medulla.

medulla oblongata. The myelencephalon, or caudal region, of the brain. Respiratory rate, heart beat, salivation, and certain other visceral activities are controlled reflexly here.

melanin (Gr. *melas,* black). Black or brown pigment in the skin. It is contained within certain of the chromatophores of lower vertebrates.

meninges, sing. meninx (Gr. *meninx,* membrane). The connective tissue membranes that ensheath the central nervous system. Mammals have three: dura mater, arachnoid, and pia mater.

mesencephalon (L. *mes-,* middle + Gr. *enkephalos,* brain). The middle, or third, of five brain regions. It lies between the diencephalon and metencephalon and includes the colliculi (mammals) or optic lobes (frogs).

mesentery (L. *mes-,* middle + Gr. *enteron,* intestine). (1) Any of the peritoneal membranes that extend from the body wall to the abdominal viscera or between visceral organs. (2) The particular membrane supporting the small intestine.

mesorchium (L. *mes-,* middle, + Gr. *orchis,* testicle). The mesentery supporting the testis.

mesovarium (L. *mes-,* middle + *ovum,* egg). The mesentery supporting the ovary.

metacarpals (Gr. *meta,* beside, after + *karpos,* wrist). Long bones in the palm of the hand that are located between the carpals and phalanges.

metatarsals (Gr. *meta,* beside, after + *tarsos,* ankle). Long bones in the foot that are located between the tarsals and phalanges.

metencephalon (Gr. *meta-,* beside, after + *enkephalos,* brain). The fourth of the five brain regions. It lies between the mesencephalon and myelencephalon and includes the cerebellum and (in mammals) the pons.

molar tooth (L. *molaris,* millstone). One of several teeth in the back of the mammalian jaw that is adapted in most cases for crushing or grinding.

monocyte (Gr. *monos,* single + *kytos,* hollow vessel, cell). A leukocyte with a kidney-shaped nucleus; a precursor of tissue macrophages.

mouth (Old English, *muth,* mouth). The cranial opening of the digestive tract; also the cavity (oral cavity) into which this opening leads.

mucosa (L. *mucus,* mucus). The lining of the digestive and respiratory tracts, many of whose cells secrete mucus.

It consists of the epithelial lining, associated glands and connective tissue, and sometimes a thin layer of smooth muscle.

muscle fiber (L. *musculus*, little mouse, muscle because of the mouselike shape of some muscles). The elongated, contractile muscle cell.

myelencephalon (Gr. *myelos*, marrow, spinal cord + *enkephalos*, brain). The most caudal of the five regions of the brain. It consists of the medulla oblongata.

mylohyoid (Gr. *myle*, millstone). A nearly transverse sheet of muscle extending between the two mandibles and the hyoid.

myo- (Gr. *mys*, gen. *myos*, like a mouse, muscle). Prefix meaning muscle or musclelike.

myoepithelium. Specialized epithelial cells containing contractile elements. Some surround the secretory cells of sweat glands and help to discharge their products.

myofibrils (Gr. *fibrilla*, small fiber). Minute, longitudinal fibrils within a muscle fiber that are barely visible with the light microscope. They are the contractile elements.

myofilaments (L. *filum*, thread). Ultramicroscopic filaments of actin and myosin that are components of the myofibrils.

naris (L. *naris*, external nostril). An external nostril.

nasal (L. *nasus*, nose). Pertaining to the nose; e.g., nasal bone, nasal cavity.

nasal conchae (L. *concha*, shell). Folds within the nasal cavity of a mammal that increase the surface area.

nasal meatus (L. *meatus*, passage). An air passage in the mammalian nasal cavity between the nasal conchae or the conchae and the nasal septum.

nephron (Gr. *nephros*, kidney). The kidney tubule, including Bowman's capsule.

nerve (L. *nervus*, sinew, nerve). A bundle of neuron processes and their investing connective tissues that extend from the brain and spinal cord to the peripheral organs.

neuron (Gr. *neuron*, sinew, nerve). A single nerve cell. It consists of dendrites, a cell body, and an axon.

neutrophile (L. *neuter*, neither + Gr. *philos*, having an affinity for). A leukocyte with minute granules in the cytoplasm that stain very lightly with acidic and basic stains. Neutrophiles are phagocytic and are the most common leukocytes.

nictitating membrane (L. *nictare*, to wink). A membrane in the median corner of the eye of many terrestrial vertebrates that can slide across the surface of the eyeball. It consists of only a vestigal semilunar fold in human beings.

nostril (Old English *nosu*, nose + *thyrl*, hole). The opening into a nasal cavity from the body surface (external nostril) or pharynx (internal nostril).

obturator foramen (L. *obturare*, p.p., *obturatus*, to close by stopping up). Large foramen in the mammalian pelvic girdle that permits the bulging of pelvic muscles arising from its periphery and from a membrane that covers it.

occipital (L. *occiput*, back of the skull). Pertains to the back of the skull; e.g., occipital bone, occipital condyles.

occipital condyle. One of a pair of condyles on the occipital bone on each side of the foramen magnum. They articulate with the cranial articular surface of the atlas.

ocular (L. *oculus*, eye). Pertaining to the eye; e.g., extrinsic ocular muscles.

oculomotor nerve. The third cranial nerve. It carries motor fibers to most of the extrinsic ocular muscles that move the eyeball.

olecranon (Gr. *olekranon*, elbow tip). The proximal end of the mammalian ulna, which extends behind the elbow joint.

olfactory nerve (L. *olfacere*, to smell). The first cranial nerve, which returns sensory fibers from the nose to the olfactory bulb of the brain.

omentum (L. *omentum*, membrane). One of two mesenteries that attach to the stomach. The greater omentum of a mammal is a sac-like fold passing between the body wall and the stomach; the lesser omentum extends from the stomach and duodenum to the liver.

omo- (Gr. *omos*, shoulder). A root referring to the shoulder. It is used in combination with other terms; e.g., omotransversarius muscle.

oocyte (Gr. *oion*, egg + *kytos*, hollow vessel, cell). An early stage in the development of the egg cell. The first meiotic division of a primary oocyte gives rise to a secondary oocyte and a polar body.

oogonium (Gr. *oion*, egg + *gonos*, progeny). A very early stage in the development of the egg cell. It enlarges to become a primary oocyte.

ootid (Gr. *oion*, egg + *-oeidos*, resemblance to). The final stage in the development of the egg cell. It results from the second meiotic division.

optic (Gr. *optikos*, sight). Pertaining to the eye.

optic chiasma (Gr. *chiasma*, cross). The complete (lower vertebrates) or partial (mammals) crossing of fibers in the optic nerves just anterior to the hypothalamus.

optic disc (Gr. *diskos*, disc). A disc-shaped area on the retina to which the optic nerve attaches. It lacks photoreceptive rods and cones; hence is a "blind spot."

optic lobe. The dorsal part of the mesencephalon in lower vertebrates. It is a major center for integration of sensory information and initiation of motor responses.

optic nerve. The second cranial nerve, which carries sensory fibers from the retina to the brain.

optic tract. The neuron tract leading from the optic chiasma to the thalamus or optic lobes or both.

oral cavity (L. *os*, gen. *oris*, mouth). The mouth cavity; also known as the buccal cavity.

orbit (L. *orbis*, circle). Circular cavity on the lateral surface of the skull that lodges the eyeball.

origin of a muscle (L. *origin*, beginning). That attachment of a muscle to a bone or other element that moves the lesser distance when the muscle contracts. Usually it is the proximal end of a limb muscle.

osteon (Gr. *osteon*, bone). A cylindrical, microscopic unit of bone that consists of concentric layers of bone matrix surrounding a cavity containing blood and lymph vessels; also called a Haversian system.

ostium (L. *ostium*, river mouth, opening). The entrance into certain organs; e.g., the ostium of the uterine tube.

otic capsule (Gr. *otikos*, pertaining to the ear). The part of the skull surrounding the inner ear. In mammals it appears as a lateral butress in the floor of the cranial cavity.

otolith (Gr. *otikos*, pertaining to the ear + *lithos*, stone). Calcareous granules within the vestibular sacs of the inner ear. They rest on receptive cells and their movements detect changes in the position of the body in space.

ovarian follicle (L. *ovum*, egg + *folliculus*, little bag). A group of cells within the ovary that surrounds the developing egg. It is also an endocrine gland whose primary hormone is estrogen.

ovary (L. *ovum*, egg). The female organ in which the egg cells develop and hormones are produced; e.g., estrogens and progesterone.

oviduct (L. *ovum*, egg). The passage in the females of lower vertebrates that transports eggs from the coelom to the cloaca. In mammals, it differentiates during embryonic development into the uterine tube, uterus, and part of the vagina.

ovisac (L. *ovum*, egg). An enlargement at the caudal end of the oviduct in certain lower vertebrates where eggs accumulate before their discharge and fertilization.

ovulation (L. *ovulum*, little egg). The discharge of eggs from the ovarian follicles and ovary into the coelom from which they enter the oviduct or uterine tube.

palate (L. *palatum*, palate). The bony roof of the oral cavity (hard palate) separating in mammals the oral and nasal cavities, and the fleshy extension of the hard palate in mammals that separates the nasal and oral pharynx (soft palate).

pampiniform plexus (L. *pampinus*, tendril + *forma*, shape). A network of veins entwining the testicular artery as it approaches the testis. It helps to reduce the temperature of blood going to the testis because heat flows from the artery to the cooler blood in the veins.

pancreas (Gr. *pan*, all + *kreas*, flesh). A large gland attached to the duodenum that secretes enzymes that act on all categories of food. It also contains the hormone-producing islets of Langerhans.

papilla amphibiorum. The part of the amphibian inner ear that is receptive to low-frequency sound waves. It is located between the utriculus and sacculus.

paraflocculi (Gr. *para*, beside + L. *floccus*, tuft of wool). Small lobes on the lateral surface of the cerebellum in certain mammals. They are conspicuous in the rat.

parasympathetic (Gr. *para*, beside + *patheticos*, sensitive). Pertaining to the parasympathetic part of the autonomic nervous system. Parasympathetic stimulation promotes digestive and other "vegetative" processes and inhibits processes activated by the sympathetic system.

parathyroid gland (Gr. *para*, beside). One of several small endocrine glands that usually are embedded on the deep surface of the thyroid gland. It secretes a hormone essential for calcium and phosphorous metabolism.

parietal (L. *paries*, wall of a room). Pertaining to the wall of the body; e.g., parietal peritoneum and parietal pleura.

parotid gland (Gr. *para*, beside + *otikos*, pertaining to the ear). Large mammalian salivary gland that lies ventral to the external ear.

patella (L. *patella*, little plate, dish). The mammalian knee cap. It lies in the tendon of the large muscle on the cranial surface of the thigh (the quadruceps femoris) and facilitates the movement of the tendon across the knee joint.

pectoral (L. *pectus*, chest). Pertaining to the chest; e.g., pectoral girdle, pectoral muscles.

pelvis (L. *pelvis*, basin). A basin-shaped structure; e.g., pelvic girdle, renal pelvis.

penis (L. *penis*, penis). The copulatory organ in the males of higher vertebrates.

pericardium (Gr. *peri-*, around + *kardia*, heart). Coelomic epithelium and associated connective tissue that covers the surface of the heart (visceral pericardium) and forms the wall of the pericardial cavity (parietal pericardium).

periosteum (Gr. *peri-*, around + *osteon*, bone). A layer of connective tissue that covers the surface of bones and serves as an attachment for tendons.

peritoneum (Gr. *peri-*, around + *tonus*, something stretched). Layer of coelomic epithelium and associated connective tissue of mesodermal origin that covers the abdominal organs (visceral peritoneum) and lines the body wall (parietal peritoneum).

peroneus muscle (Gr. *perone*, pin, fibula). One or more muscles located on the lateral surface of the lower leg and extending into the foot.

pharynx (Gr. *pharynx*, pharynx). That part of the digestive tract that lies between the mouth cavity and the esophagus. Embryonically, gill pouches develop as outgrowths of its lateral walls.

phrenic (Gr. *phren*, diaphragm). Pertaining to the diaphragm; e.g., phrenic artery and phrenic nerve.

pia mater (L. *pia*, tender + *mater*, mother). The innermost meningeal layer. It is a delicate vascular, connective tissue on the surface of the brain and spinal cord.

pineal gland (L. *pinus*, pine tree). A cone-shaped endocrine gland in mammals that is attached to the epithalamus and is thought to influence sexual development. It has evolved from the light-sensitive pineal eye, or epiphysis, of lower vertebrates.

piriformis muscle (L. *pirum*, pear). A small, triangular muscle located on the medial surface of the proximal end of the thigh.

pituitary gland. See *hypophysis.*

placenta (Gr. *plakoenta*, accusative of *plakoeis*, flat cake). Mammalian organ in which food, gases, and waste products are exchanged between the mother and fetus. It is formed by the fetal chorion and allantois and the maternal uterine lining.

plantaris longus (L. *planta*, sole of the foot). A large muscle on the caudal surface of the crus in lower terrestrial vertebrates. It attaches by the tendon of Achilles to the calcaneus and extends the foot.

platelets (Gr. *platys*, broad, flat). Small granularlike pieces of the cytoplasm of megakaryocytes in the mammalian blood stream. They contain thromboplastin, which initiates blood clots.

platysma muscle (Gr. *playts*, broad, flat). Flat, thin sheet of muscle lying just beneath the skin of the neck in mammals.

pleura (Gr. *pleura*, side, rib). Layer of coelomic epithelium and associated connective tissue that covers the lungs (visceral pleura) and lines the pleural cavities (parietal pleura).

pleuroperitoneal cavity. A combination of the peritoneal and potential pleural cavities in lower vertebrates. These species lack a diaphragm and in some cases also lack lungs.

plexus (L. *plexus*, network). A network of blood vessels, nerves, or other structures; e.g., brachial plexus, choroid plexus, pampiniform plexus.

polar body (L. *polaris*, pole). A small body located near the animal pole of a developing egg cell. Polar bodies result from an unequal separation of the cytoplasm during the first and second meiotic divisions; the nuclear material of which they are composed is then discarded.

pons (L. *pons*, bridge). The ventral part of the mammalian metencephalon where cerebral impulses are relayed to the cerebellum. Its most conspicuous surface feature is a bridgelike tract of transverse fibers.

popliteal fossa (L. *poples*, gen. *poplitis*, knee joint). The depression behind the mammalian knee joint.

portal vein (L. *portare*, to carry). A vein that carries blood from one organ to another rather than to the heart; e.g., the hepatic portal vein.

posterior (L. comparative of *posterus*, coming after). A direction toward the back surface of a human being; sometimes also used for the tail end of a quadruped, but caudal is a more appropriate term.

posterior chamber. The space within the eyeball between the iris and the lens. It is filled with aqueous humor.

premolar tooth (L. *pre-*, in front of + *molaris*, millstone). One of two or three teeth in mammals that lie in front of the molar teeth and behind the canine. They are usually adapted for a combination of cutting and grinding.

prepuce (L. *preputium*, foreskin). The foreskin of a mammal covering the glans penis.

processus vaginalis (L. *processus*, a part projecting from an organ + *vagina*, sheath). A coelomic sac that descends with the testis in mammals and ensheathes it.

prostate (Gr. *prostates*, one who stands before). A gland in male mammals located near the junction of the ductus deferens and urethra (it "stands before" the urinary bladder). It secretes much of the seminal fluid.

protraction (Gr. *pro-*, in front of + L. *trahere*, p.p. *tractus*, to pull). Muscle action that moves the entire appendage of a quadruped forward.

proximal (L. *proximus*, nearest). The end of a structure nearest its origin.

pterygoideus muscle (Gr. *pterygion*, wing). A muscle that arises from the pterygoid bone (frog) or pterygoid process on the ventral side of the skull (mammals) and inserts on the median side of the mandible. It helps to close the jaws.

pubis (L. *puber*, young adult). The cranioventral bone of the pelvic girdle to which the external genitals attach in mammals. It is covered by the pubic hair, which develops in young adults.

pudendal (L. *pudendum*, external genital region, from *pudere*, to be ashamed). Pertaining to the region of the external genitals; e.g., pudendal artery.

pulmonary (L. *pulmo*, lung). Pertaining to the lung.

pulmonary trunk. The mammalian arterial trunk; it leaves the right ventricle and soon divides into the two pulmonary arteries, which continue to the lungs.

pulmonary valve. A set of three semilunar-shaped folds in the base of the pulmonary trunk that prevents a backflow of blood into the right ventricle.

pupil (L. *pupilla*, small doll). The dark opening in the iris through which light enters the eye. The name derives from small reflections of oneself that can sometimes be seen when looking into another person's eyes.

pylorus (Gr. *pylorus*, gate keeper). The caudal region of the stomach. A muscular valve here closes the opening between the stomach and the duodenum.

pyramidal system. A group of neurons in mammals whose pyramid-shaped cell bodies lie in the motor cortex of the cerebral hemispheres and whose axons extend to the motor neurons of the brain and spinal cord. It is the major voluntary motor pathway.

quadrate cartilage (L. *quadrare*, p.p. *quadratus*, to make square). A cartilage at the caudal end of the upper jaw of a frog to which the mandibular cartilage of the lower jaw articulates. It represents part of the dorsal half of the first visceral arch.

radius (L. *radius*, ray, spoke). The bone of the forearm that rotates around the ulna in most terrestrial vertebrates. It lies on the lateral surface of the forearm when the palm is supine.

ramus (L. *ramus*, branch). A branch of a nerve or a blood vessel.

rectum (L. *rectus*, straight). The caudal part of the large intestine of a mammal. It develops from the dorsal part of the embryonic cloaca.

renal (L. *ren*, kidney). Pertaining to the kidney; e.g., renal artery.

restiform body (L. *restis*, rope + *form*, shape). A neuron tract that carries fibers from the medulla oblongata to the cerebellum. Most impulses carried relate to proprioception and equilibrium.

rete testis (L. *rete*, net). A network of small passages in mammals between the seminiferous tubules of the testis and the epididymis. They form a grossly visible cord in the rat.

retina (L. *reticulum*, little net). The innermost layer of the eyeball. It contains the photoreceptive rods and cones and associated neurons.

retraction (L. *re-*, backward + *trahere*, p.p. *tractus*, to pull). Muscle action that moves the entire appendage of a quadruped backward.

rhinal sulcus (Gr. *rhis*, gen. *rhinos*, nose). The furrow that separates the rhinencephalon from other parts of the telencephalon.

rhinencephalon (Gr. *rhis*, gen. *rhinos*, nose + *enkephalos*, brain). That part of a cerebral hemisphere related primarily to the sense of smell. It is located ventral to the rhinal sulcus in mammals.

root (Old English, *rot*, root). The origin of an organ; e.g., the dorsal and ventral roots of a spinal nerve.

rostral (L. *rostrum*, beak). A direction toward the beak. Essentially the term is a synonym for cranial.

rostral (superior) colliculus (L. *colliculus*, small hill). One of a pair of enlargements on the rostrodorsal surface of the mammalian mesencephalon that is comparable structurally to the optic lobes of lower vertebrates. It is a center for pupillary eye movement and certain other optic reflexes.

round ligament. A roundish cord; specifically, a connective tissue cord that crosses the broad ligament in female mammals and attaches to the body wall at the groin. It is comparable to the male gubernaculum.

sacculus (L. *sacculus*, little bag). The ventralmost chamber of the inner ear. It contains an otolith and functions as an equilibrium receptor.

sacrum (L. *os sacrum*, sacred bone). The group of vertebrae to which the pelvic girdle attaches in mammals. This part of an animal was favored for sacrifices in ancient times.

sagittal (L. *sagitta*, arrow). A plane of the body that passes through the sagittal suture (between the parietal bones of the skull); i.e., a median, longitudinal plane passing from dorsal to ventral.

saliva (L. *saliva*, saliva). The watery and mucous secretion of several large glands that discharge into the mouth cavity. In mammals, it contains salivary amylase (ptyalin), which initiates the chemical breakdown of starch.

sarcolemma (Gr. *sarx*, gen. *sarcos*, flesh + *lemma*, bark, sheath). The outer membrane of a muscle fiber.

sarcoplasm (Gr. *sarx*, gen. *sarcos*, flesh + *plasma*, form, mold). The cytoplasm of a muscle fiber.

sartorius muscle (Gr. *sartor*, tailor). A narrow, diagonal muscle band lying superficially on the medial surface of the thigh. It helps to cross the legs in a position assumed by early tailors.

scapula (L. *scapula*, shoulder blade). The shoulder blade, or part of the pectoral girdle that extends dorsally onto the back.

sclera (Gr. *skleros*, hard). The opaque, outermost layer of the median part of the eyeball. It is part of the fibrous tunic and is usually white.

scrotum (L. *scrotum*, pouch). The pouch of a mammal containing the testes.

sebaceous gland (L. *sebum*, tallow). A gland of mammalian skin that produces an oily secretion. It usually discharges into a hair follicle.

semicircular duct. One of three semicircular-shaped ducts of the inner ear that lie at right angles to each other and detect changes in angular acceleration (turns of the head). They are usually located in similarly shaped bony spaces in the otic capsule that are called the semicircular canals.

seminal vesicle (L. *semen*, seed + *vesicula*, blister). A gland in male mammals that is located near the union of the ductus deferens and urethra. It secretes some of the seminal fluid.

seminiferous tubule (L. *semen*, seed + *ferre*, to produce or carry). Tubules within the testis in which sperm are produced.

septum pellucidum (L. *pellucidis*, from *pelucere*, to shine through). A thin septum of nervous tissue ventral to the corpus callosum. It forms the median wall of a lateral ventricle.

serosa (L. *serum*, watery fluid, serum). The epithelial and connective tissue membrane lining body cavities and covering visceral organs. Peritoneum, pericardium, and pleura are specific serous membranes.

serratus (L. *serra*, saw). Refers to an organ with a serrated, or saw-tooth, border; e.g., the serratus ventralis muscle.

Sertoli cell. One of many large cells in the seminiferous tubules with which maturing sperm are associated.

sinus venosus (L. *sinus*, hollow). The first chamber of the heart in lower vertebrates, which receives blood from the body and leads to the right atrium in frogs. It is absorbed into the right atrium in mammals.

somatic (Gr. *soma*, body). Pertaining to the body wall rather than the internal organs; e.g., somatic muscles, somatic skeleton.

sperma- (Gr. *sperma*, gen. *spermatos*, seed, sperm). A root pertaining to the sperm.

spermatic cord. The combination in mammals of the ductus deferens and blood vessels and nerves supplying the testis. It lies within the processus vaginalis and is covered by coelomic epithelium in the fetal pig and in the rat.

spermatid. A stage in the development of sperm that follows the second meiotic division of the secondary spermatocyte. Spermatids undergo a metamorphosis to become motile spermatozoa.

spermatocyte (Gr. *kytos*, hollow vessel, cell). An early stage in the development of sperm. Primary spermatocytes undergo the first meiotic division and become secondary spermatocytes.

spermatogonium (Gr. *gonos*, progeny). The first stage in the development of the sperm cell. Spermatogonia divide mitotically to produce more spermatogonia, and some of the progeny enlarge to become primary spermatocytes.

spermatozoon (Gr. *zoon*, animal). The final, motile stage in the development of sperm.

spinal (L. *spina*, thorn, spine). Pertaining to the spine or vertebral column; e.g., spinal cord, spinal nerve.

spinous process. The dorsal, spinelike process of the vertebral arch.

splanchnic nerve (Gr. *splanchnos*, visceral organ). A nerve extending from the sympathetic trunk to ganglia at the base of the coeliac and cranial mesenteric arteries.

spleen (Gr. *splen*, spleen). A large lymphoid organ usually located near the left side of the stomach. In various vertebrates and at different times in their life cycles, it produces, stores, or destroys blood cells.

splenius muscle (Gr. *splenion*, bandage). A thin, triangular sheet of muscles on the back of the neck of a mammal located deep to the rhomboideus capitis muscle.

stapes (L. *stapes*, stirrup). The innermost of the three auditory ossicles in mammals that transmit sound waves across the tympanic cavity. Its inner end fits into the fenestra vestibuli. It is the only auditory ossicle in lower terrestrial vertebrates.

statoacoustic nerve (Gr. *statos*, standing + *akoustikos*, pertaining to hearing). A term for the eighth cranial nerve in lower vertebrates. It returns sensory fibers from the equilibrium and hearing parts of the inner ear.

sternum (Gr. *sternon*, chest). The bone on the midventral surface of the chest. Costal cartilages attach to it in mammals.

stomach (Gr. *stomakhos*, gullet, stomach). Sac-like pouch of the digestive tract, located between the esophagus and intestines, in which food is stored and the chemical breakdown of protein is initiated.

styloid process (Gr. *stylos*, pillar). A long, slender process on the underside of the skull of a mammal to which a ligament from the hyoid bone attaches.

subclavian (L. *sub-*, under + *clavicula*, small key, clavicle). Pertaining to structures that lie beneath the clavicle; e.g., subclavian artery.

sublingual gland (L. *sub-*, under + *lingua*, tongue). A mammalian salivary gland located beneath the tongue.

submucosa (L. *sub-*, under + *mucus*, mucus). A layer of vascular connective tissue in the wall of the digestive or respiratory tract that lies deep to the mucosa. It does not contain glands, as does the mucosa.

sulcus (L. *sulcus*, furrow). One of the grooves on the surface of the mammalian cerebrum between gyri.

superior (L. comparative of *superus*, upper). A direction toward the head end of a human being.

suprarenal gland (L. *supra*, above + *ren*, kidney). An endocrine gland located cranial to the kidney (mammals) or on its ventral surface (frogs). Its medullary hormone helps the body adjust to stress; its cortical hormones help regulate sexual development and the metabolism of salts, minerals, carbohydrates, and proteins. Often called the adrenal gland.

sympathetic (Gr. *sym*, together, with + *pathetikos*, sensitive). Pertaining to the sympathetic part of the autonomic nervous system. Its stimulation promotes activities needed to help the body adjust to stress and inhibits digestion and other vegetative activities.

systemic arch (Gr. *sustema*, composite whole). One of a pair of arterial arches in frogs that lead from the heart and truncus arteriosus to the dorsal aorta.

talus (L. *talus*, ankle). A proximal tarsal bone that articulates with the tibia.

tapetum lucidum (Gr. *tapes*, carpet + L. *lucere*, to shine). Part of the choroid of the eyeball in some vertebrates that reflects light back onto the retina.

tarsals (Gr. *tarsos*, ankle). The bones of the ankle.

telencephalon (Gr. *tele*, far off, distant + *enkephalos*, brain). The most anterior of the five brain regions. It includes the olfactory bulbs and cerebral hemispheres.

temporal (L. *tempus*, gen. *temporis*, time). Pertaining to the temporal region of the head; so called because hair in this region is the first to become gray in human beings.

temporal fossa. A depression on the side of the mammalian skull caudal to the orbit and dorsal to the zygomatic arch. It lodges the temporal muscle.

tendon (L. *tendere*, to stretch). A band of dense connective tissue attaching muscles to bone or to other muscles.

tendon of Achilles (Achilles, the hero of Homer's *Iliad*, was said to be invulnerable except for this tendon). The tendon that extends from the large muscle mass on the caudal surface of the crus to the calcaneus. These muscles are strong extensors of the foot.

tentorium (L. *tendere*, p.p. *tentus*, to stretch). A septum of dura mater in mammals located between the cerebrum and cerebellum. It is ossified in some species.

teres (L. *teres*, round, smooth). A term used to describe a round-shaped structure; e.g., the teres major muscle.

testis (L. *testis*, a witness, originally limited to adult males). The male organ in which sperm are produced.

thalamus (Gr. *thalamos*, inner chamber). The lateral walls of the diencephalon. It is an important sorting and relay center for sensory impulses travelling to the cerebrum.

theca (L. *theca*, case, sheath). A case or covering; e.g., the group of cells that forms the outer part of an ovarian follicle.

thoracolumbar fascia. A layer of deep fascia on the dorsal surface of the thorax and lumbar region that is associated with the latissimus dorsi and other back muscles.

thorax (Gr. *thorax*, the entire chest). The part of the mammalian body encased by ribs.

thymus (Gr. *thymos*, thymus). An aggregation of lymphoid tissue in the ventral part of the neck and thoracic cavity. It is particularly well developed in fetal and young mammals where it participates in the development of the T-lymphocyte part of the immune system.

thyroid gland (Gr. *thyreos*, oblong shield). An endocrine gland that is usually located near the cranial end of the trachea but lies over the thyroid cartilage of the larynx in human beings (hence its name). Its hormone regulates the general level of metabolism.

tibia (L. *tibia*, shinbone). The large bone on the median side of the lower leg.

tongue (Old English, *tunge*, tongue). A muscular organ in the floor of the mouth cavity in terrestrial vertebrates that is used to catch food (frogs) or to manipulate food in the mouth and to assist in swallowing the food (mammals).

trachea (Gr. *arteria trakheia*, rough artery). The main windpipe in mammals. It extends caudally from the larynx and terminates in bronchi that enter the lungs.

tract (L. *tractus*, drawing out). (1) A linear group of organs that have a common function; e.g., the digestive tract. (2) A group of neuron processes interconnecting areas within the central nervous system.

transverse (L. *trans-*, across + *vertere*, to turn). A plane of the body crossing its longitudinal axis at a right angle.

transverse process. A process of a vertebra that lies in the transverse plane of the body. In vertebral regions in which ribs are present, the tubercle of a rib articulates with it; in other regions, a small embryonic rib is incorporated in the process.

trapezoid body. An acoustic commissure at the anterior end of the ventral surface of the mammalian medulla oblongata.

trigeminal nerve (L. *tri*, three + *geminus*, twin-born). The fifth cranial nerve. It has three main branches that carry motor fibers to the jaw muscles and return sensory fibers from receptors in most of the skin over the head and the mouth lining.

trochanter (Gr. *trokhante*, from *trekhein*, to run). One of two or three processes on the proximal end of the mammalian femur to which certain pelvic muscles attach.

trochlear nerve (L. *trochlea*, pulley). The fourth cranial nerve. It carries motor fibers to an extrinsic ocular muscle.

truncus arteriosus. One of two arterial trunks in a frog that lead from the front of the heart to arterial arches supplying (1) the lungs and skin, (2) the head, and (3) the main part of the body.

tuber cinerium (L. *tuber*, bump + *cinereus*, the color of ashes). The region of the hypothalamus to which the hypophysis attaches.

tunica albuginea (L. *albus*, white). The fibrous capsule forming the wall of the testis and sending septa into the organ.

tympanic (L. *tympanum*, drum). Pertaining to the ear.

tympanic bulla (Gr. *bulla*, bubble). The bubble-shaped bony floor of the tympanic cavity in some mammals, including the rat.

tympanic cavity. The middle ear cavity.

tympanum. The eardrum.

ulna (L. *ulna*, elbow). The bone of the antebrachium that lies along the median side when the palm is supine. Its proximal end forms the olecranon, or bony projection of the elbow.

umbilical (L. *umbilicus*, navel). Pertaining to the navel; e.g., umbilical cord, umbilical artery.

ureter (L. *urina*, urine). The mammalian duct that transports urine from a kidney to the urinary bladder.

urethra (L. *urina*, urine). The mammalian duct that extends from the urinary bladder to or near the body surface. It carries only urine in females, but its distal part carries urine and sperm in males.

urinary bladder (L. *urina*, urine). A saccular organ in terrestrial vertebrates in which urine from the kidneys accumulates before discharge from the body.

urostyle (Gr. *oura*, tail + *stylos*, pillar). The spikelike caudal part of the frog's vertebral column. It consists of two fused caudal vertebrae and serves as a point of origin for certain muscles used in jumping.

uterine tube (L. *uterus*, womb). One of a pair of narrow tubes that extend from the vicinity of the ovary to the uterus. It receives the eggs at ovulation; also called fallopian tube.

uterus (L. *uterus*, womb). The organ in female mammals and certain other vertebrates in which the embryos develop.

utriculus (L. *utriculus*, small leather bottle). The chamber of the inner ear to which the semicircular ducts attach.

vagina (L. *vagina*, sheath). The part of the mammalian female reproductive tract that receives the penis during copulation.

vaginal process. See *processus vaginalis.*

vaginal vestibule (L. *vestibulum*, entrance hall). The area in a female mammal that receives the vagina and urethra. It is very shallow in human beings, but forms a short canal in the pig and rat.

vagus nerve (L. *vagus*, wandering). The tenth cranial nerve, which carries motor fibers to muscles of the larynx and parasympathetic fibers to thoracic and abdominal viscera; it returns sensory fibers from these organs.

vascular tunic (L. *vasculum*, little vessel + *tunica*, loose garment). A vascular connective tissue forming the middle layer of the eyeball wall. The choroid, ciliary body, and iris are parts of this layer.

vastus (L. *vastus*, large area). The term is used for several large muscles on the cranial surface of the thigh; e.g., vastus lateralis muscle.

vein (L. *vena*, vein). A blood vessel that carries blood toward the heart. Usually, this blood is low in oxygen content, but it may be high, as in the pulmonary vein returning blood from a lung to the heart.

vena cava (L. *vena*, vein + *cavea*, cave, hollow). One or more large veins in terrestrial vertebrates that return blood from the body and enter the sinus venosus (frog) or the right atrium of the heart (mammals).

ventral (L. *venter*, belly). A direction toward the belly surface of a quadruped.

ventricle (L. *ventriculus*, little belly). (1) A chamber of the mammalian heart that pumps blood to the lungs (right ventricle) or to the body (left ventricle). Frogs have a single ventricle, but the two blood streams remain partially separated. (2) One of the large chambers within the brain.

vertebra (L. *vertebra,* vertebra). A bone of the spinal column.

vestibulochoclea nerve (L. *vestibulum,* small entrance hall + Gr. *kokholos,* land snail). The eighth cranial nerve in mammals. It returns sensory fibers from the equilibrium (vestibular apparatus) and hearing (cochlea) parts of the inner ear.

vibrissae (L. *vibrare,* to agitate). Long, tactile hairs on the snouts of many mammals.

villi (L. *villus,* shaggy hair). Minute projections from the mucosa of the mammalian small intestine that increase its surface area.

visceral (L. *viscus,* pl. *viscera,* entrail). Pertaining to the inner part of the body rather than the body wall; e.g., visceral muscle, visceral skeleton.

vitreous body (L. *vitrum,* glass). A viscous mass of material that fills the eyeball between the retina and lens. It helps to hold the lens and retina in place and to refract light rays.

vocal cord (L. *vocalis,* speaking). A pair of folds in the lateral wall of the larynx that vibrate as air crosses them and produce vocal sounds.

vomer bone (L. *vomer,* plowshare). A plowshare-shaped bone in the roof of the mouth in frogs or in the floor of the nasal cavities in mammals.

vomeronasal organ. An accessory olfactory organ in the nasal cavity in many terrestrial vertebrates that usually connects with the mouth cavity.

vulva (L. *vulva,* covering). The external genital organs of a female mammal.

white matter (L. *materia,* matter). The whitish material (in life) of the brain and spinal cord. It is composed of the myelinated processes of neurons.

Wolffian duct. See *archinephric duct.*

zonule fibers (Gr. *zone,* girdle). Delicate fibers extending between the ciliary body and lens equator. They transmit forces from the ciliary body to the lens.

zygomatic arch (Gr. *zygon,* yolk). The arch of bone beneath the orbit of the mammalian skull that joins the facial and cranial regions.